Group Theory for the Standard Model of Particle Physics and Beyond

Series in High Energy Physics, Cosmology, and Gravitation

Series Editors: **Brian Foster**, *Oxford University, UK*
Edward W Kolb, *Fermi National Accelerator Laboratory, USA*

This series of books covers all aspects of theoretical and experimental high energy physics, cosmology and gravitation and the interface between them. In recent years the fields of particle physics and astrophysics have become increasingly interdependent and the aim of this series is to provide a library of books to meet the needs of students and researchers in these fields.

Other recent books in the series:

Particle and Astroparticle Physics
Utpal Sakar

Joint Evolution of Black Holes and Galaxies
M Colpi, V Gorini, F Haardt, and U Moschella *(Eds.)*

Gravitation: From the Hubble Length to the Planck Length
I Ciufolini, E Coccia, V Gorini, R Peron, and N Vittorio *(Eds.)*

Neutrino Physics
K Zuber

The Galactic Black Hole: Lectures on General Relativity and Astrophysics
H Falcke, and F Hehl *(Eds.)*

The Mathematical Theory of Cosmic Strings: Cosmic Strings in the Wire Approximation
M R Anderson

Geometry and Physics of Branes
U Bruzzo, V Gorini and, U Moschella *(Eds.)*

Modern Cosmology
S Bonometto, V Gorini and, U Moschella *(Eds.)*

Gravitation and Gauge Symmetries
M Blagojevic

Gravitational Waves
I Ciufolini, V Gorini, U Moschella, and P Fré *(Eds.)*

Classical and Quantum Black Holes
P Fré, V Gorini, G Magli, and U Moschella *(Eds.)*

Pulsars as Astrophysical Laboratories for Nuclear and Particle Physics
F Weber

Series in High Energy Physics, Cosmology, and Gravitation

Group Theory for the Standard Model of Particle Physics and Beyond

Ken J. Barnes

University of Southampton
School of Physics & Astronomy
United Kingdom

Taylor & Francis
6000 Broken Sound Parkway NW, Suite 300
Boca Raton, FL 33487-2742

Printed in the United States of America on acid-free paper
10 9 8 7 6 5 4 3 2 1

International Standard Book Number: 978-1-4200-7874-9 (Hardback)
DOI: 10.1201/9781439895207

Library of Congress Cataloging-in-Publication Data

Barnes, Ken J., 1938-
Group theory for the standard model of particle physics and beyond / Ken J. Barnes.
p. cm. -- (Series in high energy physics, cosmology, and gravitation)
Includes bibliographical references and index.
ISBN 978-1-4200-7879-4
1. Group theory. 2. Quantum theory. 3. Particle range (Nuclear physics) I. Title.

QC174.17.G7B37 2010
539.7'25--dc22 2009021685

Visit the Taylor & Francis Web site at
http://www.taylorandfrancis.com

and the CRC Press Web site at
http://www.crcpress.com

Contents

Preface

This book emerged out of lectures to first year postgraduate students at the then Department of Physics and Astronomy, University of Southampton, before I retired. It is hoped that this book will be appropriate for similar groups of readers in many other institutions across the world. Experimenters in this subject would probably gain much from reading this book, although some may find it difficult.

Acknowledgments

This book could never have been written without the consistently excellent help of Mrs. Hannah Williams, who handled LaTeX, figures, and packaging apparently with ease. My son, Dr. Geoffrey W. Morton, is also thanked for some of the figures and general advice. Dr. Jason Hamilton-Charlton is thanked for his generosity in providing both LaTeX and English electronic copies of my supersymmetry notes. Finally, I thank my wife, Jacky, for her continual support and help when writing anything seemed quite impossible.

Introduction

This book is definitely not a book on mathematics. It is a book on the use of symmetries, mainly described by the techniques of Lie groups and Lie algebras. Although no proofs of theorems and the like are given, except in special cases, the ideas are very firmly based on a lifetime of lecturing experience.

Introduction

1

Symmetries and Conservation Laws

You may already be familiar with the ideas of conserved quantities, such as charge in electromagnetism, but it will not hurt to go through this once more, and there may be students for whom it is quite new. Since we are dealing with elementary particles, we may as well think of conserved numbers carried on particles, and indeed we will start with the charge e on the proton. If we consider the charge of the electron ($-e$), which carries electric currents, what do we mean by "it is conserved" and what consequences might this have? We might as well, for simplicity, start with the problem in classical physics and turn to quantum mechanics later. Well, the first thing is that it cannot simply vanish or appear. Of course it can vanish by having equal but opposite charges annihilate it (producing, for example, the photons of light), or it can appear in the reverse of this. All other conserved quantities such as energy, and linear and angular momentum must be conserved—in our picture carried on the photons. Already we see that this must happen at the same time and at the same spatial point, but this is natural when the charges are carried on the particles.

You may well be familiar with the idea of conservation of charge being associated with the four divergence of the current carrying that charge. Calling j^μ the current carried by an electron (of charge $(-)e$) we can write

$$\partial_\mu j^\mu = 0\,. \tag{1.1}$$

Then we have

$$\partial\rho\partial t + \underline{\nabla}.\underline{j} = 0 \tag{1.2}$$

where ρ is the time component of j^μ and $\underline{j}$ is the spatial part of this current. If we integrate over a fixed volume we find

$$\frac{\partial\rho}{\partial t} + \text{ flow of current normally into the volume}$$
$$- \text{ flow of current normally out of the volume} = 0. \tag{1.3}$$

This means that the rate of increase of charge in the volume is equal to the rate of flow of charge into the volume minus the rate of flow out of the volume. A very natural feature of the model we use is where the charges are carried on the

DOI: 10.1201/9781439895207-1

particles. Of course, this concept needs slight changing in the world of special relativity where there is apparent contraction of lengths and dilation of times in different reference frames. Similarly in quantum mechanics further modifications are needed, which are yet further changed in quantum field theory. But we are getting too far ahead of ourselves. Let us ask what symmetries have to do with these conservation laws as our title of this chapter suggests. There is a theorem by E. Noether [1] to the effect that this is precisely what happens. It is not appropriate to prove this theorem at this stage, but it is very powerful and extends to all types of description of the physics discussed earlier. (Students note that Noether was a woman doing important work of this type at a time when there were nowhere near as many women working in science.)

The point that is necessary to understand at this stage is that all conserved quantities in physics are linked to symmetries in this way. We shall meet examples of this later. The mathematics underlying this structure is that of group theory, both discrete groups and continuous groups as described by Lie. But for the moment we move on to simple examples in the next two chapters.

Lagrangian and Hamiltonian Mechanics

Although it has been made clear that the reader is expected to be competent in quantum field theory, an exception is made at this point to be sure that the readers really can cope.

It is one of those curious quirks of history that long before quantum theory was developed this version of classical mechanics established a framework that was capable of treating both fields and particles in both classical and quantum aspects. You are strongly urged to read Chapter 19 of Volume II of *The Feynman Lectures on Physics* [2] as an introduction to the deep and fascinating approach to physics in terms of the "principle of least action," if you have not met it previously. We shall approach the topic in a more pedestrian manner than Feynman, partly because I am not so brave a teacher and partly because I want to get you calculating for yourself as soon as possible. It is my firm belief that the best way to get on top of a subject like this is to lose your fear of it by getting your hands dirty and actually doing the real calculations in detail yourself.

Suppose we have a one-dimensional system—yes, it is going to be the harmonic oscillator. We shall call the displacement from equilibrium $q(t)$ rather than $x(t)$ because later on we shall want "displacements of the fields" at various points x and we do not wish to confuse the "displacements" with the spatial positions. Then Newton's second law is replaced by the Euler–Lagrange [3] equation

$$\frac{d}{dt}\frac{\partial L}{\partial \dot{q}} = \frac{\partial L}{\partial q} \tag{1.4}$$

where $\dot{q}$ is the time derivative of q. The Lagrangian, L, is the difference between the kinetic energy (T) and the potential energy (V), that is,

$$L(q, \dot{q}) = T - V \tag{1.5}$$

and is to be regarded as a function of the independent variables q and $\dot{q}$ for the purposes of partial differentiation. For the harmonic oscillator with mass m and spring constant k we have

$$V = \frac{k}{2}q^2 = \frac{m\omega^2}{2}q^2 \tag{1.6}$$

where $\omega^2 = \frac{k}{m}$. So that

$$L = \frac{m}{2}\dot{q}^2 - \frac{m\omega^2}{2}q^2 \tag{1.7}$$

and the Euler–Lagrange equation yields

$$\frac{d}{dt}(m\dot{q}) = -m\omega^2 q \tag{1.8}$$

and we retrieve

$$\ddot{q} = -\omega^2 q \tag{1.9}$$

as expected.

Now that we have a little experience with this formalism, we can take a look at the principle of least action. You will have noticed perhaps that the concept of force (which was primary in Newton's approach) has become secondary to the idea of potential. The least action principle makes the equation of motion itself something that is derived from the minimization of the action

$$S = \int_{t_i}^{t_f} L(q, \dot{q})dt \tag{1.10}$$

where t_i and t_f are initial and final times. The principle postulates that the actual path (often alternatively called trajectory) followed by the particle is that which minimizes S. Imagine that, given L as an explicit function of q and $\dot{q}$, you evaluate S for a few paths. These are just fictitious paths and none of them is likely to be the Newtonian one. I have drawn the three from the problem on the q–t diagram in Figure 1.1.

These must start and finish at the same places and times. According to the principle, only if one of these coincides with the Newtonian path will the value of S be the minimum possible. You need a calculus approach to get a general answer. Notice, however, S is a function of the function $q(t)$. We say it is a "functional" of $q(t)$. We need to find the particular function, $q_0(t)$, that minimizes S.

Suppose there is a small variation $\delta q(t)$ in a path $q(t)$ from $q(t_i)$ to $q(t_f)$. When $q(t) = q_0(t)$, the variation δS caused by this change δq must vanish.

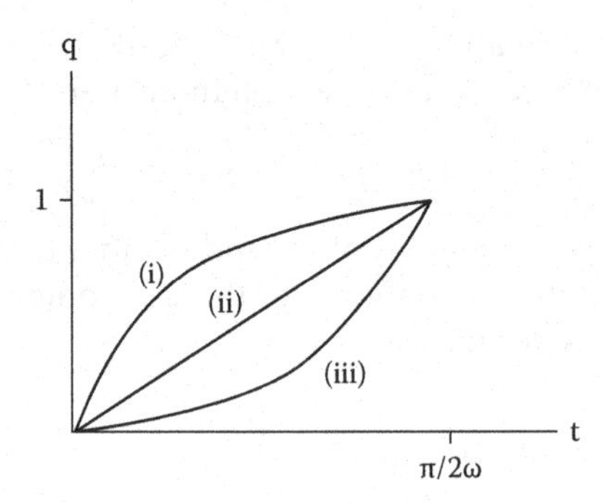

FIGURE 1.1
q–t diagram.

Now we can work out the change of action for any path as

$$\begin{aligned}\delta S &= \int_{t_i}^{t_f} \left(\frac{\partial L}{\partial q}\delta q + \frac{\partial L}{\partial \dot{q}} \right) dt \\ &= \int_{t_i}^{t_f} \left(\frac{\partial L}{\partial q}\delta q + \frac{d}{dt}\left[\frac{\partial L}{\partial \dot{q}}\right] - \frac{d}{dt}\left[\frac{\partial L}{\partial \dot{q}}\right]\delta q \right) dt \\ &= \int_{t_i}^{t_f} \delta q \left(\frac{\partial L}{\partial q} - \frac{d}{dt}\left[\frac{\partial L}{\partial \dot{q}}\right] \right) dt + \left[\frac{\partial L}{\partial \dot{q}}\delta q \right]_{t_i}^{t_f}\end{aligned}$$

where we used $\delta\dot{q} = \frac{d}{dt}\delta q$ in the second step. But we are considering paths with fixed end points, so that $\delta q(t_i) = 0 = \delta q(t_f)$ for any variation, and the final term vanishes. Hence, since δS must vanish for arbitrary δq, we need

$$\frac{d}{dt}\frac{\partial L}{\partial \dot{q}} = \frac{\partial L}{\partial q},$$

which retrieves the Euler–Lagrange equation of motion. The solution of this is the $q_0(t)$, which gives the path actually followed by the particle.

As we shall see later, this formalism is well suited to treat systems of the many (indeed infinitely many) linked dynamical variables found in field theories. But the transition from classical to quantum mechanics is made more transparent by considering the Hamiltonian formulation. The idea, in the first place, is to find a change in variables (from q and $\dot{q}$) which will replace the second order Euler–Lagrange equation by two linked first order equations. This piece of magic is performed by introducing

$$p = \frac{\partial L}{\partial \dot{q}} \tag{1.11}$$

as a "generalized momentum conjugate to the generalized coordinate q." (When q is a Cartesian coordinate, p will frequently be the usual linear momentum, as we shall see.) Then the Hamiltonian is introduced by the

Legendre transformation

$$H(q, p) = p\dot{q} - L(q, \dot{q}) \tag{1.12}$$

and the Euler–Lagrange equation is replaced by the pair of equations

$$\dot{q} = \frac{\partial H}{\partial p} \tag{1.13}$$

$$\dot{p} = -\frac{\partial H}{\partial q}, \tag{1.14}$$

which are known as Hamilton's canonical equations. To get a feel for this formulation we return to our old friend the harmonic oscillator. From Equation (1.7) we see that

$$p = \frac{\partial L}{\partial \dot{q}} = m\dot{q}, \tag{1.15}$$

which is reassuring, and we can then see that from Equation (1.12)

$$\begin{aligned} H &= \frac{p^2}{m} - \left\{ \frac{p^2}{2m} - \frac{m\omega^2}{2} q^2 \right\} \\ &= \frac{p^2}{2m} + \frac{m\omega^2}{2} q^2 \end{aligned}$$

is the form of the Hamiltonian in the new variables. Notice that the Hamiltonian is the total energy, $T + V$. This is a very general feature, and provided that time does not appear explicitly then

$$\frac{dH}{dt} = \frac{\partial H}{\partial q}\dot{q} + \frac{\partial H}{\partial p}\dot{p} = \frac{\partial H}{\partial q}\frac{\partial H}{\partial p} + \frac{\partial H}{\partial p}\left[-\frac{\partial H}{\partial q} \right] = 0, \tag{1.16}$$

which reflects energy conservation. In the present case the equations of motion, Equations (1.13) and (1.14), yield

$$\dot{q} = \frac{p}{m} \tag{1.17}$$

$$\dot{p} = -m\omega^2 q \tag{1.18}$$

when Equation (1.16) is used directly. The first of these reconfirms the definition of the momentum, and on substitution into the second retrieves Equation (1.9) as the second order equation of motion. It turns out, however, to be instructive to solve the first order Equations (1.17) and (1.18) directly. Consider the linear combination

$$A = \frac{1}{\sqrt{2}} \left(x\sqrt{m\omega} + ip\frac{1}{\sqrt{m\omega}} \right), \tag{1.19}$$

which is so designed that

$$\dot{A} = -i\omega A \tag{1.20}$$

$$A = ae^{-i\omega t} \tag{1.21}$$

as the obvious solution, where a is constant. Taking the complex conjugate of Equation (1.19), we immediately find

$$x = \frac{1}{\sqrt{2m\omega}}(A + A^*) = \frac{1}{\sqrt{2m\omega}}(ae^{-i\omega t} + a^*e^{i\omega t}), \tag{1.22}$$

which is equivalent to the previous solution.

Quantum Mechanics

The passage to quantum mechanics in this formalism is facilitated by introducing the Poisson bracket notation. The Poisson bracket of any two functions f and g, of q and p, is simply

$$\{f, g\} \equiv \frac{\partial f}{\partial q}\frac{\partial g}{\partial p} - \frac{\partial f}{\partial p}\frac{\partial g}{\partial q} \tag{1.23}$$

and we see that

$$\{q, H\} = \dot{q} \tag{1.24}$$

$$\{p, H\} = \dot{p} \tag{1.25}$$

are alternative ways of writing Equations (1.13) and (1.14), the equations of motion. Moreover, if F is any function of q and p, then

$$\begin{aligned} \frac{dF}{dt} &= \frac{\partial F}{\partial q}\dot{q} + \frac{\partial F}{\partial p}\dot{p} \\ &= \frac{\partial F}{\partial q}\{q, H\} + \frac{\partial F}{\partial p}\{p, H\} = \{F, H\} \end{aligned} \tag{1.26}$$

while

$$\begin{aligned} \{q, q\} &= 0 \\ \{p, p\} &= 0 \\ \{q, p\} &= 1 \end{aligned} \tag{1.27}$$

follow directly from the definition (Equation (1.24)) of the Poisson bracket. The transition to quantum mechanics is now effected by the correspondence $\{\alpha, \beta\} \rightarrow -i[\hat{\alpha}, \hat{\beta}] = -i(\hat{\alpha}\hat{\beta} - \hat{\beta}\hat{\alpha})$ between the classical dynamical variables and their hatted quantum mechanical operator correspondences. (We use "natural" units with $\hbar = 1$.) In particular, Equation (1.27) yields

$$[\hat{q}(t), \hat{p}(t)] = i \tag{1.28}$$

expressing the Heisenberg uncertainty principle [5], and Equation (1.26) gives

$$\frac{d\hat{F}(t)}{dt} = -i[\hat{F}(t), H] \tag{1.29}$$

as the Heisenberg equation of motion. The time dependence has been exhibited to draw the reader's attention to the fact that this is quantum mechanics expressed in the Heisenberg picture [6], where states are time independent but the dynamical variables contain the time dependence.

The alternative Schrödinger picture, in which the variables are time independent, has the time dependence of state vectors given by the Schrödinger equation

$$\hat{H}|\psi(t) > = i\frac{\partial}{\partial t}|\psi(t) > \tag{1.30}$$

with the formal solution

$$|\psi(t) > = e^{-i\hat{H}}|\psi > \tag{1.31}$$

where we have identified the Schrödinger state at time zero with $|\psi(0) >$, with $|\psi >$ the time independent Heisenberg state. Of course, Equation (1.31) is just a unitary transformation between the two pictures, with

$$\hat{F}(t) + e^{i\hat{H}}(t) = e^{i\hat{H}}(t)\hat{F}e^{-i\hat{H}}(t) \tag{1.32}$$

as the corresponding transformation between operators. The important feature of this is that

$$[\hat{q}, \hat{p}] = i \tag{1.33}$$

follows immediately from Equation (1.28) as an expression of the uncertainty principle in the Schrödinger picture. In quantum field theory we shall find the Heisenberg picture very convenient.

In the quantum case we have the operator version of Equation (1.15)

$$\hat{H} = \frac{\hat{p}^2}{2m} + \frac{m\omega^2}{2}\hat{q}^2 \tag{1.34}$$

$$= \frac{\hat{p}^2(t)}{2m} + \frac{m\omega^2}{2}\hat{q}^2(t) \tag{1.35}$$

with Equation (1.32) giving trivially the equality of these alternate forms. From the Heisenberg equation of motion (Equation (1.29)) we can easily see that

$$\dot{\hat{q}}(t) = \frac{\hat{p}(t)}{m} \tag{1.36}$$

$$\dot{\hat{p}}(t) = -m\omega^2 \dot{\hat{q}}(t) \tag{1.37}$$

so that we get

$$\ddot{\hat{q}}(t) = -m\omega^2 \hat{q}(t) \tag{1.38}$$

by combining these. Now, please notice that this is not just the classical equation of motion (Equation (1.9)) again. What Equation (1.38) tells us is the behavior of the operator with time, not where the particle can be found. If we take the expectation value of Equation (1.38) between (time independent) Heisenberg states, then we learn that the mean position of the particle does follow the classical path. This is very reassuring, but there will be quantum fluctuations about the classical path, of course.

The Oscillator Spectrum: Creation and Annihilation Operators

This subtopic is of such central importance later that it deserves a section all to itself. You have no doubt all been exposed to this material before, but I want to stress the operator treatment that we shall see again in our field theory. (If you already know this method, it will at least serve as a review and to establish notation.)

We seek a set of states $|E_n>, n = 0, 1, \ldots,$ to serve as a complete basis in which to expand any general state, and thus must solve the time independent Schrödinger equation

$$\hat{H}|E_n> E_n|E_n> \tag{1.39}$$

for the eigenvalues and eigenvectors. The Hamiltonian is given in Equation (1.34) as

$$\hat{H} = \frac{\hat{p}^2}{2m} + \frac{m\omega^2}{2}\hat{q}^2$$

but our classical treatment suggests Equation (1.19)

$$\hat{a} = \frac{1}{\sqrt{2}}\left(\hat{q}\sqrt{m\omega} + i\hat{p}\frac{1}{\sqrt{m\omega}}\right) \tag{1.40}$$

$$\hat{a}^\dagger = \frac{1}{\sqrt{2}}\left(\hat{q}\sqrt{m\omega} - i\hat{p}\frac{1}{\sqrt{m\omega}}\right) \tag{1.41}$$

as preferable dynamical variables. It is straightforward to see that

$$\begin{aligned}\hat{a}^{\dagger}\hat{a} &= \frac{m\omega}{2}\hat{q}^2 + \frac{1}{2m\omega}\hat{p}^2 + \frac{i}{2}[\hat{q},\hat{p}] \\ &= \frac{1}{\omega}\left(\frac{\hat{p}^2}{2m} + \frac{m\omega^2}{2}\hat{q}^2\right) - \frac{1}{2}\end{aligned}$$

where Equation (1.33) is used in the last term. Hence we have

$$\hat{H} = \omega\hat{a}^{\dagger}\hat{a} + \frac{\omega}{2} \tag{1.42}$$

$$\hat{H} = \omega\hat{a}\hat{a}^{\dagger} - \frac{\omega}{2} \tag{1.43}$$

so that

$$\hat{H} = \frac{\omega}{2}(\hat{a}^{\dagger}\hat{a} + \hat{a}\hat{a}^{\dagger}) \tag{1.44}$$

$$[\hat{a},\hat{a}^{\dagger}] = 1 \tag{1.45}$$

follow by adding and subtracting. Notice that (from Equation (1.42)),

$$[\hat{H},\hat{a}^{\dagger}] = \omega\hat{a}^{\dagger}[\hat{a},\hat{a}^{\dagger}]$$

$$[\hat{H},\hat{a}^{\dagger}] = \omega\hat{a}^{\dagger} \tag{1.46}$$

$$[\hat{H},\hat{a}] = -\omega\hat{a}. \tag{1.47}$$

(In Equations (1.45)–(1.47) we now have the algebraic information in a suitable form to find the spectrum. I urge you to do Problem 1.14 before continuing.) We are now in a position to see exactly why $\hat{a}$ and $\hat{a}^{\dagger}$ are so important. They have the magical property in that they take you from one energy eigenstate into another, rather than into some arbitrary linear combination of states. To see this, consider the effect of the Hamiltonian on an eigenstate that has been changed by the action of $\hat{a}^{\dagger}$

$$\begin{aligned}\hat{H}\hat{a}^{\dagger}|E_n> &= ([\hat{H},\hat{a}^{\dagger}] + \hat{a}^{\dagger}\hat{H})|E_n> \\ &= (\omega\hat{a}^{\dagger} + \hat{a}^{\dagger}E_n)|E_n> \\ &= (E_n + \omega)\hat{a}^{\dagger}|E_n>\end{aligned} \tag{1.48}$$

so we see that $\hat{a}^{\dagger}|E_n>$ is indeed an eigenstate of $\hat{H}$ and $(E_n+\omega)$ is the eigenvalue. In a similar way we can establish that $\hat{a}|E_n>$ is an eigenstate with $(E_n-\omega)$ as the eigenvalue this time. Of course, you cannot lower the energy until it becomes negative, so there must be a ground state of lowest energy E_0 with

$$\hat{a}|E_0> = 0 \tag{1.49}$$

as its definition to maintain consistency. (Beware! In relativistic physics such reasoning will not be true.) But here you can prove it. From Equation (1.42) we see that

$$\hat{H}|E_0> = 0 + \frac{1}{2}\omega|E_0>$$

establishing $E_0 = \frac{1}{2}\omega$ as the ground state (or zero point) energy. Then, by raising, we see that the energy spectrum is

$$E_n = \left(n + \frac{1}{2}\right)\omega \qquad n = 0, 1, \ldots \tag{1.50}$$

and the corresponding eigenstates are given by

$$|E_n> = \frac{(\hat{a}^\dagger)^n}{\sqrt{n!}}|E_0> \tag{1.51}$$

where the exact factor follows from the requirement

$$< E_n|E_n> = 1 \tag{1.52}$$

of normalization. It is now natural to speak of a vacuum rather than a ground state, and then to envisage the "creation of particles" (or "excitation of quanta") into that vacuum. Indeed if we define a number operator

$$\hat{N} = \hat{a}^\dagger\hat{a} \tag{1.53}$$

to conform to our notation in Equations (1.42) and (1.50), then the change of notation to

$$\hat{N}|n> = n|n> \tag{1.54}$$

$$\hat{H}|n> = E_n|n>= (n + 1/2)\omega|n> \tag{1.55}$$

becomes irresistible.

Coupled Oscillators: Normal Modes

Before we launch into an attack on the quantum field theory of infinitely many degrees of freedom, it is probably sensible to try a finite number of variables. Let's start with the classical theory of two equal masses in a one-dimensional space (e.g., in a straight slot on a horizontal table) tied together by a spring of spring constant g, and tied to fixed points by springs of spring constant k.

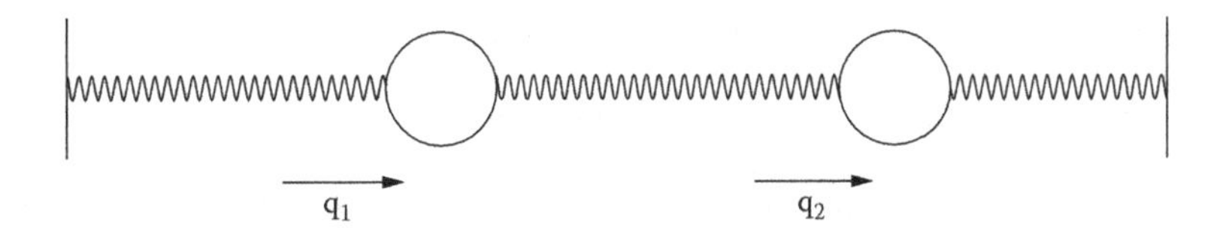

FIGURE 1.2
Three spring forces.

I have in mind the picture in Figure 1.2, where q_1 and q_2 are displacements from equilibrium, and the Lagrangian takes the form

$$L = \frac{1}{2}m(\dot{q}_1^2 + \dot{q}_2^2) - \frac{1}{2}k(\dot{q}_1^2 + \dot{q}_2^2) - \frac{1}{2}g(q_2 - q_1)^2 \tag{1.56}$$

if none of the springs are stretched or compressed in the equilibrium position. You can think of this as a model of a (very) small solid. One advantage of the Lagrangian approach is that we never have to introduce the forces in the springs and then eliminate them again; constraints are handled very neatly in this formalism. The Euler–Lagrange equations yield

$$\ddot{q}_1 = -\frac{k}{m}q_1 + \frac{g}{m}(q_2 - q_1) \tag{1.57}$$

$$\ddot{q}_2 = -\frac{k}{m}q_2 - \frac{g}{m}(q_2 - q_1), \tag{1.58}$$

which are sufficiently simple that we do not need formal methods to solve them. We spot the relevant combinations of variables by adding and subtracting to obtain

$$(\ddot{q}_1 + \ddot{q}_2) = -\frac{k}{m}(q_1 + q_2) \tag{1.59}$$

$$(\ddot{q}_2 - \ddot{q}_1) = -\left(\frac{k}{m} + \frac{2g}{m}\right)(q_2 - q_1), \tag{1.60}$$

which we recognize as uncoupled simple harmonic oscillators. The solutions are then obvious. We have one normal mode of oscillation with frequency

$$\omega_1 = \sqrt{\frac{k}{m}} \tag{1.61}$$

and Equation (1.60) is satisfied trivially by having the two displacements equal. The second normal mode has frequency

$$\omega_2 = \sqrt{\frac{k + 2g}{m}} \tag{1.62}$$

and Equation (1.59) is satisfied trivially by the two displacements being equal but opposite in sense. The general solution is then obtained by superposition (because the equations are linear) and we have

$$q_1 = Acos(\omega_1 t + \delta) + Bsin(\omega_2 t + \varepsilon) \tag{1.63}$$

$$q_2 = Acos(\omega_1 t + \delta) - Bsin(\omega_2 t + \varepsilon) \tag{1.64}$$

where A, B, δ, and ε are constants to be fixed by initial conditions.

Notice that the frequency of the lowest mode is independent of g—naturally, because if $q_2 = q_1$, then the middle spring is not stretched. Indeed, if $k = 0$ this mode is of zero frequency. You can then think of a two atom molecule that is free, and this mode corresponds to the free motion of the center of mass. This is not really important for these lectures, but when you hear theorists worrying about zero modes and symmetries you will have some idea of their problems. Zero modes can be a real pain for theorists as they usually need separate treatment, and the associated symmetry (here just translation) is not always easy to find.

Now, how do we quantize a system like this? The key lies in the observation made earlier that we are just dealing with uncoupled harmonic oscillators in terms of

$$Q_1 = \frac{1}{\sqrt{2}}(q_1 + q_2) \quad \text{and} \quad Q_2 = \frac{1}{\sqrt{2}}(q_2 - q_1) \tag{1.65}$$

as the variables, where the normalization factor is for convenience. It is now easy to work out the form

$$H = \frac{1}{2m}\left(P_1^2 + P_2^2\right) - \frac{k}{2}Q_1^2 - \left(\frac{k}{2} + g\right)Q_2^2 \tag{1.66}$$

for the Hamiltonian. So to quantize we simply have to put hats on the Q's and P's, and do the harmonic oscillator problems twice. You should check, of course, that the commutation relations are

$$[\hat{Q}_i, \hat{Q}_j] = 0$$
$$[\hat{Q}_i, \hat{P}_j] = i\delta_{ij}$$
$$[\hat{P}_i, \hat{P}_j] = 0 \tag{1.67}$$

where i,j = 1,2, as you expect in terms of the new variables. Then there is a vacuum state, two number operators $\hat{N}_1$ and $\hat{N}_2$ (one for each of $\hat{Q}_1$ and $\hat{Q}_2$ systems), and states $|n_1, n_2>$ with $(n_1 + \frac{1}{2})\omega_1 + (n_2 + \frac{1}{2})\omega_2$ as the energies.

Now it is probably fairly clear that this idea generalizes to the N variables case. The eigenstates of the Hamiltonian are denoted by $|n_1, n_2 \ldots n_N>$, and are associated with energy eigenvalues $E_{n_1}, n_2 \ldots n_N = \sum_{r=1}^{N}(n_r + \frac{1}{2})\omega_r$ with the various ω_r depending upon the details of masses and spring constants. Notice that the emphasis has now changed completely from "displacements of atoms" to "excitations in the solid." It is particularly important to note that there is no restriction (except the total energy available) on the number of excitations, even though the number of underlying "displacements" is still quite finite at this stage. This approach is central to many body physics where one speaks of "phonons" as the vibrational excitations (similarly plasmons for plasma oscillations, and magnons for magnetic oscillations). What we are leading toward is a framework in which all elementary particles (quarks, leptons, W^{+-}, Z_0, photons, gluons) are excitations of underlying fields. However, we must first learn to handle a simple classical continuous system.

One-Dimensional Fields: Waves

What is to be our generalization of the Lagrangian in Equation (1.56) when there is a continuum of "atoms" rather than just two? The sum over two sites becomes an integral over the position x, and presumably the last term becomes proportional to a spatial derivative. We shall have to absorb dimensions into the constants, of course, and we use $\phi(x,t)$ rather than $q(x,t)$ for future convenience, so that we write

$$L = \int \left[1/2 \left(\frac{\partial \phi}{\partial t} \right)^2 - \frac{\mu^2}{2}\phi^2 - \frac{c^2}{2}\left(\frac{\partial \phi}{\partial x} \right)^2 \right] \rho dx \tag{1.68}$$

where ρ is a mass density, and μ and c are just convenient names for the modified constants. Now what is the generalization of the Euler–Lagrange equations? In this continuous case we define a Lagrangian density $\mathcal{L}$ by

$$L = \int \mathcal{L} dx \tag{1.69}$$

so that

$$S = \int L dt = \int dt dx \mathcal{L} \tag{1.70}$$

is the action to be minimized. (I am being deliberately vague about the limits of the integrals. You may think of a solid between $x = 0$ and $x = L$, or of a field extending over all space.) The new feature in the continuous case is that $\mathcal{L}$ depends not only on ϕ and $\dot{\phi} = \frac{\partial \phi}{\partial t}$ but also on $\frac{\partial \phi}{\partial x}$ and all these must be varied. Thus

$$\delta S = \int dt dx \left[\frac{\partial \mathcal{L}}{\partial \phi}\delta\phi + \frac{\partial \mathcal{L}}{\partial \dot{\phi}}\delta\dot{\phi} + \frac{\partial \phi}{\partial(\partial\phi/\partial x)}\delta\left(\frac{\partial \phi}{\partial x} \right) \right] \tag{1.71}$$

and we integrate by parts in t for the middle term, and by parts in x for the final term, to get

$$\delta S = \int dtdx \left[\frac{\partial \mathcal{L}}{\partial \phi} - \frac{\partial}{\partial t}\left(\frac{\partial \mathcal{L}}{\partial \dot{\phi}}\right) - \frac{\partial}{\partial x}\left(\frac{\partial \mathcal{L}}{\partial\left(\partial \phi/\partial x\right)}\right)\right] \delta \phi \tag{1.72}$$

when the end-point (or boundary) terms are assumed to vanish. Since S is arbitrary we get the Euler–Lagrange field equations

$$\frac{\partial \mathcal{L}}{\partial \phi} = \frac{\partial}{\partial t}\left(\frac{\partial \mathcal{L}}{\partial \dot{\phi}}\right) + \frac{\partial}{\partial x}\left(\frac{\partial \mathcal{L}}{\partial(\partial \phi/\partial x)}\right) \tag{1.73}$$

with

$$\frac{\partial \mathcal{L}}{\partial \phi} = \frac{\partial}{\partial t}\left(\frac{\partial \mathcal{L}}{\partial \dot{\phi}}\right) + \underline{\nabla}\left(\frac{\partial \mathcal{L}}{\partial(\underline{\nabla}\phi)}\right) \tag{1.74}$$

as the generalization to three spatial dimensions.

Putting the expression for $\mathcal{L}$ implicit in Equation (1.68) into Equation (1.73) we get

$$0 = \frac{\partial^2 \phi}{\partial t^2} - c^2 \frac{\partial^2 \phi}{\partial x^2} + \mu^2 \phi \tag{1.75}$$

as the field equation. Now for some good news and some bad news. The good news is that (not entirely by accident) this happens to be a relativistic equation suitable for discussing spinless particles (like the Higgs boson), so our warm-up exercises are already covering relevant material. The bad news is that there are a few snags in its interpretation in relativistic physics. For the moment we merely have to notice that we already know a lot about this equation. If we ignore the μ^2 term, we have

$$\frac{\partial^2 \phi}{\partial t^2} = c^2 \frac{\partial^2 \phi}{\partial x^2}, \tag{1.76}$$

which is the familiar wave equation (from electromagnetism, e.g., where ϕ would be a component of the E or B field) with general solution

$$\phi = f(x - ct) + g(x + ct) \tag{1.77}$$

where f and g are arbitrary functions representing, respectively, waves traveling to the right and left with velocity c, which we shall now set equal to one in "natural" units. Moreover, we are familiar with the idea of superimposing sinusoidal solutions to produce standing waves, when discrete frequencies arise from the boundary conditions, as in the case of sound waves in a tube or transverse waves on a violin string. Then a general solution can be written as a Fourier series of these sinusoidal ones.

But this is already enough hint to see what we need to do for the full equation Equation (1.75)—the sinusoidal functions will still work, but the frequencies will be μ^2 dependent. Assume for simplicity that the end "atoms" in our linear chain are constrained to be at rest at $x = 0$ and $x = L$. Then a suitable trial solution ensuring this is

$$\phi_r(x, t) = A_r(t)\sin\left(\frac{r\pi x}{L}\right) \quad (1.78)$$

provided that r is an integer. Substituting into Equation 1.75 gives

$$\ddot{A}_r = -\omega_r^2 A_r \quad (1.79)$$

$$\omega_r^2 = \mu^2 + \frac{r^2\pi^2}{L^2} \quad (1.80)$$

and we see that we are back to the simple harmonic oscillator again. Of course a general solution, by superposition, may be written as

$$\phi(x, t) = \sum_{r=1}^{\infty} A_r(t) sin\left(\frac{r\pi\ x}{L}\right) \quad (1.81)$$

with each $A_r(t)$ being associated with its own characteristic r frequency as given earlier. So what we have discovered is that the Fourier series gives the mode analysis. The sinusoidal functions in x give exactly the correct "linear combinations of coordinates" to pick out the separate frequencies—the Fourier amplitudes, A_r, act exactly as do normal coordinates. To confirm this view we can construct the Hamiltonian for this system, and see if the solution (Equation 1.81) reveals a sum of uncoupled oscillators. It is clear in our expression for the Lagrangian (Equation 1.68) that the first term is a kinetic term and the remainder is potential, so that

$$H = \int \left[\frac{1}{2}\left(\frac{\partial\phi}{\partial t}\right)^2 + \frac{\mu^2}{2}\phi^2 + \frac{1}{2}\left(\frac{\partial\phi}{\partial x}\right)^2\right] \rho dx \quad (1.82)$$

is the form of the Hamiltonian. Substituting Equation (1.81) and using the orthonormality of the sine functions in the region $x = 0$ to $x = L$ reveals that

$$H = \frac{\rho L}{2}\sum_{r=1}^{\infty}\left[\frac{1}{2}\dot{A}_r^2 + \frac{1}{2}\omega_r^2 A_r^2\right] \quad (1.83)$$

confirming the view we had formed. The quantum version of this problem is now obvious—this is exactly as previously shown except that there are an infinite number of oscillators, and therefore an infinite set of the n_r to be specified as occupation numbers to define the state. This does raise, however, the question of the zero-point energy problem. We have now introduced an infinite number of oscillators each with minimum energy $\frac{1}{2}\omega$, so that the total energy of our "continuous chain of atoms" is infinite. The conventional view is that this is not a real problem, only the energy differences really matter. (After all, there is no concept of destroying this "crystal" into an infinite number of parts and trying to extract the energy.) So you subtract the infinite number of $\frac{1}{2}\omega$ contributions and take a new reference point for zero energy.

The Final Step: Lagrange–Hamilton Quantum Field Theory

We now have just about enough experience to attempt the real problem. The starting point will be some Lagrangian for a field $\phi(x, t)$. (In more physical models, later, this would be perhaps a multicomponent field, say the electromagnetic field or the electron field. The excitations of the field will be identified with particles.)

As a Lagrangian we take

$$L = \int \left[\frac{1}{2}\left(\frac{\partial \phi}{\partial t}\right)^2 - \frac{1}{2}\left(\frac{\partial \phi}{\partial x}\right)^2 - \frac{\mu^2}{2}\phi^2 \right] dx, \tag{1.84}$$

which is just Equation 1.68 with $c = 1$ and ρ absorbed into the field. (This is conventional. Notice that the dimensions of ϕ now vary with the number, d, of spatial, not time, dimensions. With $\hbar = l = c$, the only dimension is mass is proportional to $(length)^{-1}$ is proportional to $(time)^{-1}$, and then the dimension of ϕ is $\frac{1}{2}(d-1)$ to ensure that the dimension of L, and H, is unity.) To be able to quantize, we need to extend the idea of conjugate momentum so that we can work in Hamiltonian form. Noting our treatment of the Lagrangian in Equation (1.69), we introduce a Hamiltonian density $\mathcal{H}$, so that

$$H = \int \mathcal{H} dx \tag{1.85}$$

and define a "momentum field" $\pi(x, t)$ by

$$\pi(x, t) = \frac{\partial \mathcal{L}}{\partial \dot{\phi}} \tag{1.86}$$

in direct analogy to Equation (1.11). Usually this is referred to as the "momentum canonically conjugate to ϕ." Then by analogy with Equation (1.12), we write

$$\mathcal{H}(\phi, \pi) = \pi(x, t)\dot{\phi}(x, t) - \mathcal{L} \tag{1.87}$$

and see from Equation (1.84) that for our model

$$\pi(x, t) = \dot{\phi} \tag{1.88}$$

$$\mathcal{H}(\phi, \pi) = \frac{1}{2}\pi^2(x, t) + \frac{1}{2}\left(\frac{\partial \phi(x, t)}{\partial x}\right)^2 + \frac{\mu^2}{2}\phi^2(x, t), \tag{1.89}$$

which in comparison with Equation (1.82) is very encouraging.

To go to the quantized version of field theory we elevate the objects ϕ and π to operators $\hat{\phi}$ and $\hat{\pi}$. Then we have to postulate appropriate commutators between them. (Please note that this is a new and postulated idea. We are working in analogy with quantum mechanics, but this lack of commutation between $\hat{\phi}$ and $\hat{\pi}$ is not a consequence of, for example, x having become a quantum operator; on the contrary x is perfectly classical here. For this reason the field quantization is frequently referred to as "second quantization," and in the version we shall propose "canonical second quantization.") Now obviously we wish to postulate commutation relations that mimic Equation 1.67 as closely as possible in the continuous case. The generalization of the Kronecker δ_{ij} with two indices over which summation with an arbitrary vector, f, gives back the vector as

$$\sum_j f_j \delta_{ij} = f_i \tag{1.90}$$

is the Dirac delta function $\delta(x - y)$ with

$$\int_{-\infty}^{\infty} f(x)\delta(x - y)dx = f(y) \tag{1.91}$$

for all reasonable functions f, as the defining property. Again, we notice that in Equation (1.67), as previously in Equation 1.28, the commutation relations are between the time dependent operators at equal times, for example,

$$[\hat{Q}_i(t), \hat{P}_j(t)] = i\delta_{ij} \tag{1.92}$$

in the case of the most crucial commutators. We postulate, therefore, the *equal-time-commutation relations*

$$\begin{aligned} [\hat{\phi}(x,t), \hat{\phi}(y,t)] &= 0 \\ [\hat{\phi}(x,t), \hat{\pi}(y,t)] &= i\delta(x - y) \\ [\hat{\pi}(x,t), \hat{\pi}(y,t)] &= 0 \end{aligned} \tag{1.93}$$

with obvious generalization, through $\delta^3(\underline{x}-\underline{y})$, to three dimensions. Variables, such as $\hat{\phi}$ and $\hat{\pi}$, related in this way are said to be "conjugate to each other."

We now promote Equations (1.88) and (1.89) to

$$\hat{\pi}(x,t) = \dot{\hat{\phi}}(x,t) \quad \text{and} \quad \hat{\mathcal{H}} = \frac{1}{2}\hat{\pi}^2 + \frac{1}{2}\left(\frac{\partial\hat{\phi}}{\partial x}\right)^2 + \frac{\mu^2}{2}\hat{\phi}^2 \tag{1.94}$$

as operator equations. To find the eigenvalues and eigenstates of

$$\hat{H} = \int \mathcal{H} dx$$

we simply make a Fourier expansion. This time we shall not restrict ourselves to a box of length L, but let x run from $-\infty$ to $+\infty$, and use running waves instead of standing waves. Now we expand as

$$\phi(x, t) = \int_{-\infty}^{\infty} \frac{dk}{2\pi} \frac{1}{2\omega} \left[a(k)e^{ikx-i\omega t} + a^*(k)e^{-ikx+i\omega t}\right] \tag{1.95}$$

where $\omega^2 = \mu^2 + k^2$ and we have required ϕ to be real, so that

$$\pi(x, t) = \int_{-\infty}^{\infty} \frac{dk}{2\pi} \frac{1}{2\omega} \left[a(k)e^{ikx-i\omega t} + a^*(k)e^{-ikx+i\omega t}\right] \tag{1.96}$$

follows at once. (Do not worry about the conventional factors of 2π and 2ω. It turns out to be very convenient in the relativistic interpretation. Just think of them as factors, which allow a smooth comparison with wave function normalization.) When we quantize, this becomes promoted to

$$\hat{\phi}(x, t) = \int_{-\infty}^{\infty} \frac{dk}{2\pi} \frac{1}{2\omega} \left[\hat{a}(k)e^{ikx-i\omega t} + \hat{a}^\dagger(k)e^{-ikt+i\omega t}\right] \tag{1.97}$$

with corresponding expression for $\hat{\pi}$, and the crucial point is that the Fourier coefficients have become operators. It is straightforward to see that the commutation relations (Equation (1.91)) determine the commutators

$$\begin{aligned} [\hat{a}(k), \hat{a}(k')] &= 0 \\ [\hat{a}(k), \hat{a}^\dagger(k')] &= 2\pi 2\omega\delta(k - k') \\ [\hat{a}^\dagger(k)\hat{a}^\dagger(k')] &= 0 \end{aligned} \tag{1.98}$$

for the mode operators.

You can see that $\hat{a}^\dagger$ and $\hat{a}$ are almost certainly creation and annihilation operators for this continuum case. We need only substitute our expansions for $\hat{\phi}$ and $\hat{\pi}$ into Equation (1.95) to see that

$$\hat{H} = \int \frac{dk}{2\pi} \frac{1}{2\omega} \frac{\omega}{2} \left[\hat{a}^\dagger(k)\hat{a}(k) + \hat{a}(k)\hat{a}^\dagger(k)\right] \tag{1.99}$$

to confirm our hopes. If we rewrite this as

$$\hat{H} = \int \frac{dk}{2\pi} \frac{1}{2\omega} \omega\hat{a}^\dagger(k)\hat{a}(k) + \int \frac{dk}{2\pi} \frac{1}{2\omega} \frac{\omega}{2} [\hat{a}^(k), \hat{a}^\dagger(k)] \tag{1.100}$$

then using Equation (1.99) we recognize the second term as the infinite sum of zero point energies, which we have learned to discard. Thus we can take

$$\hat{H} = \int \frac{dk}{2\pi} \frac{1}{2\omega} \omega\hat{a}^\dagger(k)\hat{a}(k) \tag{1.101}$$

and the spectrum will obviously follow along the usual simple harmonic oscillator lines. There will be a ground state, or vacuum, $|0>$, determined by

$$\hat{a}^{\dagger}(k)|0> = 0 \text{ for all k prime} \tag{1.102}$$

into which $\hat{a}^{\dagger}(k)$ will create a quantum of frequency ω in the now familiar way.

Notice the interpretation of the ground state as a vacuum with no particles in it. This is central to the interpretation of modern quantum field theory in elementary particle physics.

To get a little more feel for this new structure, consider the amplitude of $\hat{\phi}$ between the vacuum and the one particle state of momentum p:

$$\begin{aligned} < 0|\hat{\phi}(x,t)|p> &=< 0|\hat{\phi}(x,t)\hat{a}^{\dagger}(p)|0> \\ &=< 0|\int \frac{dk}{2\pi}\frac{1}{2\omega}[\hat{a}(k)e^{ikx-i\omega t} + \hat{a}^{\dagger}(k)e^{-ikx+i\omega t}]\hat{a}^{\dagger}(p)|0> . \end{aligned} \tag{1.103}$$

From the conjugate of Equation (1.103) we see that the second term gives zero, and with this trick in mind we rewrite the first term as

$$< 0|\hat{\phi}(x,t)|p> =< 0|\int \frac{dk}{2\pi}\frac{1}{2\omega}e^{ikx-i\omega t}[\hat{a}(k), \hat{a}^{\dagger}(p)]|0> \tag{1.104}$$

to enable us to use Equation (1.99). Thus we find

$$< 0|\hat{\phi}(x,t)|p> =< 0|\int \frac{dk}{2\pi}\frac{1}{2\omega}e^{ikx-i\omega t}2\pi 2\omega\delta(k-p)|0> \tag{1.105}$$

$$= e^{ipx-i\omega t} \tag{1.106}$$

where $\omega^2 = \mu^2 + p^2$ now, and $< 0|0 >= 1$ has been assumed. You will probably recognize this as the wave function for this problem. (We actually looked at standing waves earlier, which are superpositions of these. But the $\mu^2 = 0$ case should be very familiar.) So this is where the old wave functions of quantum mechanics appear; they are vacuum to one-particle matrix elements of the field operators.

Finally, think about the two-particle state

$$|k_1, k_2> = \hat{a}^{\dagger}(k_1)\hat{a}^{\dagger}(k_2)|0> \ . \tag{1.107}$$

Because of Equation (1.99), we see that this is symmetric under the $k_1 \leftrightarrow k_2$ interchange. There is no way to distinguish one quantum of energy (particle) from another—we must be dealing with bosons. Obviously, something will have to be modified later to handle fermions—and then the spin—and interactions between particles.

References

1. E. Noether, *Hachr. Akad. Wess. Götingerm Math.-Phys. KP. II* (1918): 235; M.A. Tevel. *Transport Theory Statist. Phys.* 1, no. 3 (1971): 183.
2. R.P. Feynman, *The Feynman Lectures on Physics*, Vol. 2. Addison-Wesley, Reading, MA, 1964, chapter 19.
3. H. Goldstein. Euler–Lagrange equation. *Classical Mechanics*, 2nd ed. Addison-Wesley, Reading, MA, 1980, p. 44.
4. J. Poisson, Poisson Bracket, *J. de l'École Polytech* 8, (1809); 266, Whittaker, 1944, p. 299.
5. W. Heisenberg, Heisenberg: The Uncertainty Relations. *Zeitschrift für Physik* 43 (1927): 172.
6. G. Sterman. Heisenberg Picture, Appendix A. *An Introduction to Quantum Field Theory*. Cambridge University Press, 1993.
7. G. Sterman. Schrödinger Picture, Appendix A. *An Introduction to Quantum Field Theory*. Cambridge University Press, 1993.
8. G. Artken. Fourier Series. In *Mathematical Methods for Physicists*, 3rd ed., 1985, chapter 14.
9. N. Highan. Kronecker delta. *Handbook of Writing for the Mathematical Sciences*, Society for Industrial and Applied Mathematics, 1998.
10. P.A.M. Dirac. *Quantum Mechanics*, 4th ed. Oxford University Press, London.

Problems

1.1 Solve $\ddot{x} = -\omega^2 x$ to find $x = x_{max} cos(\omega t + \delta)$ where δ and x_{max} are constants.

1.2 Write down the Lagrangian for a body of mass m experiencing the acceleration g due to gravity. Hence, solve the problem of a body dropped from rest. (Yes, it really is trivial.)

1.3 For the harmonic oscillator, evaluate S for the three paths (a) $q = sin\omega t$ (this is the Newtonian path, it should give minimum S); (b) $q = at$; and (c) $q = bt^2$ adjusting a and b to ensure the same endpoint for all three paths.

1.4 Write out dH to check that the Legendre transformation really does yield Hamilton's canonical equations.

1.5 Show that if the kinetic term has the form

$$T = \sum_{i,j} \frac{1}{2} m_{ij} \dot{q}_i \dot{q}_j$$

where i and j label the N particles of a system, and the m_{ij} are generalized "masses" independent of the velocities $\dot{q}_i$, then $H = T + V$ whenever the potential is velocity independent. (These circumstances do arise whenever the constraint equations [from the variables you first use to the q_i] are independent of time. Try writing the kinetic term for a single particle in two dimensions first in Cartesian and then in polar coordinates.)

1.6 Check that Hamilton's equations for Problem 1.2 do yield the same second order equation on substitution. Solve the first order equations directly, and confirm the previous result.

1.7 Start from $A = \alpha x + \beta p$ where α and β are constants and "design" the form in Equation 1.19 for yourself.

1.8 Show that the two solutions are equivalent and find the relationships between the constants of the two solutions.

1.9 Check that the Poisson bracket equations of motion give the usual results for the particle falling under gravity.

1.10 Check that matrix elements of $\hat{F}(t)$ between Heisenberg states agree with those of $\hat{F}$ between Schrödinger states. (Yes, these questions are trivial.)

1.11 Derive Equation (1.33) from Equation (1.28).

1.12 Derive Equations (1.36) and (1.37) from Equation (1.29).

1.13 If you are not sure about the meaning of Hermitian operators (like $\hat{q}$ and $\hat{p}$) please look up the idea. As a check, show that $\hat{p} \rightarrow -i\frac{\partial}{\partial \hat{q}}$ (in the Schrödinger representation) is Hermitian.

1.14 The operator solution of the harmonic oscillator is of central importance. To make sure that you have got the idea up to this point, close your notes and work it again starting with $\hat{H} = \hat{x}^2 + \hat{p}^2$ so that the constants are different.

1.15 Go on - do it!

1.16 Take the expectation value of $\hat{H} = \hat{x}^2 + \hat{p}^2$ for an arbitrary state, and use the hermiticity of $\hat{x}$ and $\hat{p}$ to show that you have a sum of moduli squared, hence, not negative.

1.17 Assume $C_n|E_{n+1} >= \hat{a}^\dagger|E_n >$, where $< E_n|E_n >= 1$. Now use $< E_{n+1}|E_{n+1} >= 1$ to find C_n is real. Now show that Equation (1.51) is correct. Find the wave function for the first excited state, explicitly. Hint: First use Equation (1.49) to find the ground state wave function.

1.18 Derive Equations (1.57) and (1.58).

1.19 This is a question you can ignore if you like. Try three equal masses in a line joined only by springs (of constant g) between the middle one and each of the end ones and otherwise free. You should get three equations of motion. Try solutions in which all three masses have a single frequency (normal mode, of course), to get three algebraic questions. Find the three values of the frequency that make these (homogeneous) equations compatible. (You can put the determinant at zero. Alternatively, pick out the zero frequency mode and the problem looks easy enough to guess the configurations of the other modes.)

1.20 Please work out Equation (1.66) for yourself.

1.21 Please check Equation (1.67). Start by the usual commutators for the original variables and remember that $[\hat{q}_1, \hat{p}_2] = 0$ and so forth.

1.22 What will the wave function look like? Write the wave function explicitly for $n_1 = 0 = n_2$.

1.23 If this is not clear to you, try Problem 1.19 then think about Problem 1.22 if there are three masses.

1.24 Go on - do it!

1.25 If you feel weak, just verify that Equation (1.77) solves Equation (1.76). If you feel strong try to prove that this is the general solution. (Change variables to $x \pm ct$.)

1.26 Try superimposing two sine waves each of wavelength λ and frequency ν but traveling in opposite directions. If ν_0 is the lowest frequency mode on a pipe of length L open at one end and closed at the other, what is the speed of sound?

1.27 Go on - do it! $[cos2A - cos2B = 2sin(A + B)sin(B - A).]$

1.28 Check this please.

1.29 Check again please.

1.30 Check that the dimensions work out for $[\hat{\phi}(\underline{x}, t)\hat{\pi}(\underline{y}, t) = i\delta^d(\underline{x} - \underline{y})]$ in d space dimensions.

1.31 Actually it is not quite so straightforward—you need to be able to invert the Fourier transform. But you can easily verify that Equations (1.98) and (1.99) do give Equation (1.93), so please do this.

1.32 Go on - it gets a bit messy - but you can do it!

1.33 Show this, please. By now you really should be able to solve the harmonic oscillator by the operator method!

2

Quantum Angular Momentum

Index Notation

Index notation is the modern and easy way to work through problems such as we now face. It has always seemed to the author that many good, even great, physicists have never learned this topic and therefore find sections of books and papers that do use it very hard or even impossible to follow. It is truly easy to learn and takes only a little practice to become competent in its use. Parts of this section will appear again later, sometimes as problems, so the equations here will be numbered I1, I2, and so forth to emphasize the point.

Indices in this section are lowercase letters that can be attached as subscripts or superscripts to appropriate things, such as momentum or angular momentum. An example might be J_i where J is an Hermitian $J^\dagger = J$ angular momentum operator and i is a subscript in the range (1, 2, 3) or (x, y, z) specifying which direction of component is being treated. Such an index, appearing once and once only, is called a *free* index. It is free for you to pick within the range specified. Such indices must *balance* in equations. For example,

$$A_i = 7B_i + 12C_i \tag{I1}$$

would be acceptable whereas $A_i = 7B$ or $A_i = 7B_i + 12C$ is not. An index repeated once and only once in an expression is called a *dummy* index and implies summation. This is known as the Einstein convention. For example,

$$A_i A_i \equiv A_1 A_1 + A_2 A_2 + A_3 A_3 \tag{I2}$$

where you must take great care not to confuse the power A_i squared, that is, $(A_i)^2$ with the second component $(A_i)^2$. At this stage indices can appear either as subscripts or superscripts. The safe way to write things is to reserve numbers for powers (or use brackets). There is a mathematical theorem to the effect that there are two numerically invariant (i.e., not changing under, say,

DOI: 10.1201/9781439895207-2

a rotation even if A_i does) tensors. The first is the Kronecker delta δ_{ij}, which is symmetric in the two indices and has the values

$$\delta_{ij} = 0 \quad \text{if} \quad i \neq j \tag{I3}$$

$$\delta_{ij} = 1 \quad \text{if} \quad i = j. \tag{I4}$$

This has three important implications.

$$\delta_{ij} A_j = A_i, \tag{I5}$$

which can be seen by writing out the sum as

$$\delta_{i1} A_1 + \delta_{i2} A_2 + \delta_{i3} A_3, \tag{I6}$$

which then yields, for example,

$$1 \times A_1 + 0 \times A_2 + 0 \times A_3 = A \tag{I7}$$

if $i = 1$ and so forth

Second,

$$\delta_{ii} = 3, \tag{I8}$$

which can be confirmed by writing out the sum as

$$\delta_{11} + \delta_{22} + \delta_{33} = 1 + 1 + 1 = 3 \tag{I9}$$

in the three dimensions in which we are working.

Third,

$$\delta_{ij} \delta_{jm} = \delta_{im}, \tag{I10}$$

which can be seen by expanding the sum to read

$$\delta_{i1} \delta_{1m} + \delta_{i2} \delta_{2m} + \delta_{i3} \delta_{3m}. \tag{I11}$$

The second numerically invariant tensor is the Levi-Civita tensor ε_{ijk}, which is totally antisymmetric in the indices with $\varepsilon_{123} = 1$ by convention. Obviously $\varepsilon_{123} = 1$ tells us that $\varepsilon_{231} = 1$ and so forth but also $\varepsilon_{132} = (-)1$ and so forth and $\varepsilon_{112} = 0$.

Quantum Angular Momentum

We define an angular momentum by a set of three operators, $J_i, i = 1, 2, 3$ or x, y, z, which satisfy

$$[J_i, J_j] = i\varepsilon_{ijk} J_k \tag{2.1}$$

$$J_i^\dagger = J_i \tag{2.2}$$

where $\hbar \equiv 1$ in our natural unit, and the Einstein summation convention has been used. This simply means that an index repeated once is summed over, as distinct from those appearing once, which is a free one for you to pick. The same index appearing three or more times is an error.

Here ε_{ijk} is the Levi-Civita tensor (density), which is antisymmetric in any pair of indices with $\varepsilon_{123} = 1$.

Now the existence of this (Lie) algebra is very far from trivial and the content is very high indeed. We will look at the latter aspect first. Switch from the Cartesian basis to a spherical one by defining

$$J_\pm = J_1 \pm i J_2 \tag{2.3}$$

so that

$$[J_3, J_\pm] = \pm J_\pm \tag{2.4}$$

$$[J_+, J_-] = 2J_3 \tag{2.5}$$

with

$$J_3^\dagger = J_3, \tag{2.6}$$

$$J_+^\dagger = J_-, \tag{2.7}$$

$$J_-^\dagger = J_+. \tag{2.8}$$

Then consider the operator

$$\underline{J}^2 = J_i J_i \tag{2.9}$$

$$= J_- J_+ + J_3(J_3 + 1) \tag{2.10}$$

$$= J_+ J_- + J_3(J_3 - 1). \tag{2.11}$$

Notice that

$$[J_i, \underline{J}^2] = [J_i, J_j J_j] \tag{2.12}$$

$$= J_j [J_i, J_j] + [J_i, J_j] J_j. \tag{2.13}$$

(You should write out [A, BC] explicitly to understand this point.)

$$= i\varepsilon_{ijk}(J_j J_k + J_k J_j) = 0 \text{ by symmetry.} \tag{2.14}$$

$$[J_i, \underline{J}^2] = 0. \tag{2.15}$$

Such an operator is called a Casimir. You can always use Casimirs as part of your complete commuting set of observables. We will take $\underline{J}^2$ and J_3—you will learn later that there can be no more—and set up the eigenvalue problem as

$$\underline{J}^2|\beta, m> = \beta|\beta, m> \tag{2.16}$$

$$J_3|\beta, m> = m|\beta, m> . \tag{2.17}$$

Then returning to our theme of the angular momentum we see that

$$J_3 J_+|\beta, m> = (J_+ J_3 + J_+)|\beta, m> \tag{2.18}$$

$$(J_+ J_3 + J_+)|\beta, m> = (m+1)J_+|\beta, m> \tag{2.19}$$

also

$$\underline{J}^2 J_+|\beta, m> = J_+\underline{J}^2|\beta, m> \tag{2.20}$$

$$|\beta, m> = \beta J_+|\beta, m> \tag{2.21}$$

so that

$$J_+|\beta, m> = c^+(\beta, m)|\beta, m+1> \tag{2.22}$$

unless $c^+(\beta, m)$ vanishes for $m_{max} = j$, say. Now use the normalization to see that

$$c^+(\beta, m)^+ c^+(\beta, m) < \beta, m+1|\beta, m+1> =< \beta, m|J_- J_+|\beta, m> \tag{2.23}$$

$$=< \beta, m|\{\underline{J}^2 - J_3(J_3+1)\}|\beta, m> \tag{2.24}$$

$$= \beta - m(m+1) \tag{2.25}$$

therefore

$$|c^+(\beta, m)|^2 = \beta - m(m+1). \tag{2.26}$$

Similarly

$$J_-|\beta, m> = c^-(\beta, m)|\beta, m-1> \tag{2.27}$$

with $|c^-(\beta, m)|^2 = \beta - m(m-1)$. Of course, β has to be positive because $\underline{J}^2$ is a sum $J_i J_i^+$ (therefore like the sum of norms of vectors). More formally

$$\beta =< \beta, m|\underline{J}^2|\beta, m > \tag{2.28}$$

$$= \sum_n < \beta, m|J_i|n >< n|J_i^\dagger|\beta > \tag{2.29}$$

$$= \sum_n | < \beta|J_i|n > |^2 \geq 0. \tag{2.30}$$

Clearly then, m cannot get too big for fixed β, and $c^+(\beta, j) = 0$ tells us

$$\beta = j(j+1). \tag{2.31}$$

Neither can m become too negative, and

$$j(j+1) = m_{min}(m_{min} - 1) \tag{2.32}$$

tells us $m_{min} = -j$ (or $j+1$, which is crazy).

Finally, the step length is an integer so we see that j is an integer or half-integer.

Result

$$\underline{J}^2|j, m> = j(j+1)|j, m> \tag{2.33}$$

$$J_3|j, m> = m|j, m> \tag{2.34}$$

$$-j \leq m \leq j \text{ with each taking integer (or half) values.} \tag{2.35}$$

$$< j'm'|J_3|jm> = m\delta_{jj'}\delta_{mm'} \text{ where } \delta_{ij} \text{ is the Kronecker delta} \tag{2.36}$$

$$< j'm'|\underline{J}^2|jm> = j(j+1)\delta_{jj'}\delta_{mm'} \tag{2.37}$$

$$\text{with} \quad \delta_{11} = \delta_{22} = \delta_{33} = 1 \tag{2.38}$$

$$\text{and} \quad \delta_{ij} = 0 \text{ if } i \neq j \tag{2.39}$$

$$< j'm'|J_\pm|j_m> = \sqrt{j(j+1) - m(m \pm 1)}\delta_{jj'}\delta_{m\pm 1,m'}. \tag{2.40}$$

The coefficient of the square root is real and positive by phase convention.

Matrix Representations

Any problem with precisely N eigenvalues and eigenvectors (all nondegenerate) may be represented on the space of

$$A_N|n> = n|n> \qquad n = 1, \ldots, N\,, \tag{2.41}$$

which has the matrix representation

$$<r|A_n|c> \Rightarrow \begin{pmatrix} 1 & 0 & \cdots & 0 \\ 0 & 2 & \cdots & 0 \\ \vdots & & \ddots & \vdots \\ 0 & \cdots & \cdots & N \end{pmatrix}, \tag{2.42}$$

$$\text{and} \quad |r> \Rightarrow \begin{pmatrix} 0 \\ 0 \\ \vdots \\ 1 \\ \vdots \\ 0 \end{pmatrix} \tag{2.43}$$

where the 1 is in the rth position.

Spin $\frac{1}{2}$

$$\underline{S}^2|\frac{1}{2}, \pm\frac{1}{2}> = \frac{1}{2}\frac{3}{2}|\frac{1}{2}, \pm\frac{1}{2}>\,. \tag{2.44}$$

$$S_3|\frac{1}{2}, \pm\frac{1}{2}> = \pm\frac{1}{2}|\frac{1}{2}, \pm\frac{1}{2}>\,. \tag{2.45}$$

This may be represented on the A_2 space, with labels taken as $\pm\frac{1}{2}$ for convenience, that is,

$$|\frac{1}{2}, \frac{1}{2} > \Rightarrow \begin{pmatrix} 1 \\ 0 \end{pmatrix} \tag{2.46}$$

$$|\frac{1}{2}, -\frac{1}{2} > \Rightarrow \begin{pmatrix} 0 \\ 1 \end{pmatrix}. \tag{2.47}$$

We call

$$S_i = (\hbar)\frac{1}{2}\Sigma_i \tag{2.48}$$

and look at the matrix representations.

$$< m|\Sigma_3|n > = 2m\delta_{mn} \Rightarrow \sigma_3 = \begin{pmatrix} 1 & 0 \\ 0 & -1 \end{pmatrix} \tag{2.49}$$

$$< m|\Sigma_+|n > = 2\sqrt{\frac{1}{2}\frac{3}{2} - m(m+1)}\delta_{n+1,m} \Rightarrow \sigma_+ = \begin{pmatrix} 0 & 2 \\ 0 & 0 \end{pmatrix} \tag{2.50}$$

$$< m|\Sigma_-|n > = \ldots \Rightarrow \sigma_- = \begin{pmatrix} 0 & 0 \\ 2 & 0 \end{pmatrix} \tag{2.51}$$

$$\sigma_x = \sigma_1 = \frac{1}{2}(\sigma_+ + \sigma_-) = \begin{pmatrix} 0 & 1 \\ 1 & 0 \end{pmatrix} \tag{2.52}$$

$$\sigma_y = \sigma_2 = \frac{1}{2i}(\sigma_+ - \sigma_-) = \begin{pmatrix} 0 & -i \\ i & 0 \end{pmatrix}. \tag{2.53}$$

Of course,

$$\underline{S}^2 \Rightarrow \frac{1}{2}\frac{3}{2}\begin{pmatrix} 1 & 0 \\ 0 & 1 \end{pmatrix} = \frac{3}{4}1. \tag{2.54}$$

The three σ_i (often called τ_i) are called Pauli matrices.

This last property is called completeness. It is the usual matrix completeness here or completeness in the Dirac sense. A proof of this might be obtained by expanding an arbitrary matrix M as

$$M = \frac{1}{\sqrt{2}}M^i\sigma_i + \frac{1}{\sqrt{2}}M^0 1 \tag{2.55}$$

where clearly

$$M^i = \frac{1}{\sqrt{2}} Tr(M\sigma^i) \tag{2.56}$$

$$M^0 = \frac{1}{\sqrt{2}} Tr(M1). \tag{2.57}$$

Then

$$M = \frac{1}{2}\sigma_i Tr(M\sigma_i) + \frac{1}{2} 1 Tr(M) \tag{2.58}$$

so that

$$2M^\beta_\alpha = (\sigma_i)^\beta_\alpha M^\gamma_\delta (\sigma^i)^\delta_\gamma + \ldots. \tag{2.59}$$

Hence we get

$$M^\gamma_\delta \left\{(\sigma_i)^\beta_\alpha (\sigma^i)^\delta_\gamma + \ldots\right\} = 0 \tag{2.60}$$

but M was quite arbitrary.

Note that a general state of spin $\frac{1}{2}$ is now written as

$$|\Psi> = a_+|\frac{1}{2}, \frac{1}{2}> + a_-|\frac{1}{2}, -\frac{1}{2}>$$

$$\left[\Rightarrow \begin{pmatrix} a_+ \\ a_- \end{pmatrix} = \chi_\alpha\right] \tag{2.61}$$

$$\text{with } |a_+|^2 + |a_-|^2 = 1 \tag{2.62}$$

and the probability of "spin-up" being given by

$$|<\frac{1}{2}, \frac{1}{2}|\Psi>|^2 = |a_+|^2. \tag{2.63}$$

Now deduce

$$\varepsilon_{ijk}\varepsilon^{lmk} = \delta^l_i \delta^m_j - \delta^m_i \delta^l_j \tag{2.64}$$

$$\text{and } \varepsilon_{ijk}\varepsilon^{ljk} = 2!\delta^l_i \tag{2.65}$$

$$\text{and } \varepsilon_{ijk}\varepsilon^{ijk} = 3!. \tag{2.66}$$

Addition of Angular Momenta

Suppose we have two quite independent systems (well separated angular momenta, or distinct elementary particles carrying isospin, etc.) so that the

FIGURE 2.1
The distribution of angular momentum.

algebra is

$$\left[j_i^A, j_j^A\right] = i\varepsilon_{ijk} j_k^A \tag{2.67}$$

$$\left[j_i^B, j_j^B\right] = i\varepsilon_{ijk} j_k^B \tag{2.68}$$

$$\left[j_i^A, j_j^B\right] = 0 \tag{2.69}$$

where A and B label the systems. We can take our complete commuting set of observables as $\{j_z^A, (\underline{j}^A)^2, j_z^B, (\underline{j}^B)^2\}$; then the states have the outer product form.

It will frequently be convenient to describe the system by an alternative set, which includes the total angular momentum

$$J_i = j_i^A + j_i^B. \tag{2.70}$$

Moreover, $(\underline{j}^A)^2$ and $(\underline{j}^B)^2$ are Casimirs. So an alternative complete commuting set of observables is $\{\underline{J}^2, J_z, (\underline{j}^A)^2, (\underline{j}^B)^2\}$ and we label the states $|J, M, j^A, j^B> \equiv |J\, M>$ frequently when the Casimir labels are dropped.

Now we need to know how the eigenvalues and their ranges are related, and also how to find the coefficients (Clebsch–Gordan) in the linear relationship between $|J\, M>$ and $|j^A m^A>|j^B m^B>$. Well

$$M = j_z^A + j_z^B \tag{2.71}$$

so you have

$$M = m^A + m^B \tag{2.72}$$

where we take $A > B$.

Again, each J will carry the usual $(2J+1) M$ value so the number of times you get M is

$$n(M) = \sum_{J \geq M} N(J) \tag{2.73}$$

where $N(J)$ is the number of times you get J. Hence,

$$N(J) = n(M = J) - n(M = J+1). \tag{2.74}$$

Now we can find $n(M)$. You get M by $m_A + m_B$ subject to the ranges. Take $j_B \leq j_A$ and draw a picture and work your way in from the extreme right. Clearly you get nothing until you get down to $j_A + j_B$ (see Figure 2.1). Thus, $n(M) = 0$ if $|M| > j_A + j_B$.

Clebsch–Gordan Coefficients

To find the precise coefficient it is most instructive simply to work an example that shows all the features.

$$1 \otimes \frac{1}{2} = \frac{3}{2} \oplus \frac{1}{2}. \tag{2.75}$$

We must relate $\left| 1, \begin{matrix} +1 \\ 0 \\ -1 \end{matrix} \right\rangle \left| \tfrac{1}{2}, \pm\tfrac{1}{2} \right\rangle$ to $\left| \tfrac{3}{2}, \begin{matrix} \pm\tfrac{3}{2} \\ \\ \pm\tfrac{1}{2} \end{matrix} \right\rangle$ and $\left| \tfrac{1}{2}, \pm\tfrac{1}{2} \right\rangle$.

Notice for speed that

$$\sqrt{j(j+1) - m(m-1)} = \sqrt{(j+m)(j+1-m)} \tag{2.76}$$

starts as $\sqrt{(2j)(1)}$ when $m = j$, then $\sqrt{(2j-1)2}$ and so on.

We have

$$|\frac{3}{2}, \frac{1}{2} > = \sqrt{\frac{2}{3}}|1, 0 > |\frac{1}{2}, \frac{1}{2} > + \sqrt{\frac{1}{3}}|1, 1 > |\frac{1}{2}, -\frac{1}{2} > \tag{2.77}$$

and can repeatedly lower to get

$$|\frac{3}{2}, -\frac{1}{2} > = \sqrt{\frac{1}{3}}|1, -1 > |\frac{1}{2}, \frac{1}{2} > + \sqrt{\frac{2}{3}}|1, 0 > |\frac{1}{2}, -\frac{1}{2} >, \tag{2.78}$$

$$\text{and} \quad |\frac{3}{2}, -\frac{3}{2} > = |1, -1 > |\frac{1}{2}, \frac{-1}{2} > . \tag{2.79}$$

Now, we need $|\frac{1}{2}, \pm\frac{1}{2} >$. Well $M = m_A + m_B$ tells us

$$|\frac{1}{2}, \frac{1}{2} > = a|11 > |\frac{1}{2}, \frac{-1}{2} > + b|1, 0 > |\frac{1}{2}, \frac{1}{2} > . \tag{2.80}$$

Then:

1. Normalization $\Rightarrow |a|^2 + |b|^2 = 1$.
2. Orthogonality to $< \frac{3}{2}, \frac{1}{2}| \Rightarrow a = -\sqrt{2}b$. (You could get this also by $J^+|\frac{1}{2}\frac{1}{2} >= 0$.)
3. Condon and Shortley phase convention (theory of atomic spectra). $\{< j_A j_A| < j_B, J - j_A|\}|J J >$ is real and > 0. ($j_A \geq j_B$) implies $a > 0$.

The coefficients relating the two sets of descriptions are called Clebsch-Gordan coefficients [1].

Hence

$$|\frac{1}{2}, \frac{1}{2} > = \sqrt{\frac{2}{3}}|1, 1 > |\frac{1}{2}, -\frac{1}{2} > - \frac{1}{\sqrt{3}}|1, 0 > |\frac{1}{2}, \frac{1}{2} > \tag{2.81}$$

and we can lower to

$$|\frac{1}{2} - \frac{1}{2} > = \sqrt{\frac{1}{3}}|1, 0 > |\frac{1}{2}, -\frac{1}{2} > -\sqrt{\frac{2}{3}}|1, -1 > |\frac{1}{2}, \frac{1}{2} > . \qquad (2.82)$$

You can now solve these equations to get

$$|1, 0 > |\frac{1}{2}, -\frac{1}{2} > = \sqrt{\frac{1}{3}}|\frac{1}{2} - \frac{1}{2} > +\sqrt{\frac{2}{3}}|\frac{3}{2}, -\frac{1}{2} > \text{ and so on.} \qquad (2.83)$$

The coefficients are called Clebsch–Gordan (or vector-addition coefficients). You get tables of them [1].

Notes

1. Can read either way.
2. If you do $j^B + j^A$ then

$$\{< j^A m^A| < j^B m^B|\}|J\,M> = (-1)^{J-j^A-j^B}\{< j^B m^B| < j^A m^A|\}|J\,M> . \qquad (2.84)$$

   which is important mainly if $j^A = j^B$ and $J - j^A - j^B$ is odd.
3. Spectroscopic notation $^{2S+1}(S, P, D, F, G \ldots for\, L)_J$ where S, P, D, F, and so forth stand for the spectroscopic series sharp, principal, diffuse, fine, and so on.
4. When adding L to $\frac{1}{2}$ they use lowercase letters and leave off the spin multiplicity superscript. For example,

$$\begin{array}{ll} j = \frac{1}{2} & s_{\frac{1}{2}} \text{ and } p_{\frac{1}{2}} \\ j = \frac{3}{2} & p_{\frac{3}{2}} \text{ and } d_{\frac{3}{2}} \end{array} .$$

5. General formalism follows from completeness:

$$\begin{aligned} |J\,M> &= \left\{\sum_{m_A m_B} |m_A m_B >< m_A m_B|\right\} |J\,M> \\ &= \sum_{m_A m_B} < m_A m_B|J\,M> |m_A m_B> \end{aligned} \qquad (2.85)$$

   where $< m_A m_B|J\,M>$ is CG - real by our conventions. Similarly,

$$\begin{aligned} |m_A m_B> &= \left(\sum_{J\,M} |J\,M>< J\,M|\right) |m_A m_B> \\ &= \sum_{J\,M} < J\,M|m_A m_B> |J\,M> \end{aligned} \qquad (2.86)$$

   which is why the tables read both ways.

6. Notation.

$$< j^A j^B m^A m^B | j^A j^B J\, M > .$$

$$C\, j_A j_B (J, M; m_A m_B).$$

$$S^{j_A j_B}_{J\, m_A m_B} .$$

Wigner "$3 - j$" symbol

$$\begin{pmatrix} j^A & j^B & J \\ m^A & m^B & -M \end{pmatrix} = \frac{(-1)^{j_A - j_B + M}}{\sqrt{2J+1}} < j^A j^B m^A m^B | j^A j^B J\, M > .$$

Matrix Representation of Direct (Outer, Kronecker) Products

We have worked with states $|m_A > |m_B >$, and used $J_i = (j_i^A \times 1) + (1 \times j_i^B)$ in a formal way. To represent this on matrices we need the direct product construction.

If ϕ and θ are $(m \times 1)$ and $(n \times 1)$ matrices (columns) you can form the direct, or outer or Kronecker product

$$(\phi \times \theta)_{i\alpha} = \phi_i \theta_\alpha = \begin{pmatrix} \phi_1\theta_1 \\ \phi_1\theta_2 \\ \vdots \\ \phi_2\theta_1 \\ \phi_2\theta_2 \\ \vdots \end{pmatrix}, \tag{2.87}$$

which is an $(mn \times 1)$ matrix. More generally, if A is $m \times m$ and B is $n \times n$ then the direct product is

$$(A \times B)_{i\alpha, j\beta} = (A_{ij} B_{\alpha\beta}) = \begin{pmatrix} a_{11}B & a_{12}B & \cdots & a_{1m}B \\ a_{21}B & \cdots & & \cdots \\ \vdots & & \ddots & \vdots \\ a_{m1}B & \cdots & & a_{mn}B \end{pmatrix}, \tag{2.88}$$

which is an $(mn \times mn)$ matrix.

$\frac{1}{2} \otimes \frac{1}{2} = 1 \oplus 0$ in Matrix Representation

Recall

$$|\phi> \Rightarrow \begin{pmatrix} \phi^{+} \\ \phi^{-} \end{pmatrix} \text{ and } |\theta> \Rightarrow \begin{pmatrix} \theta^{+} \\ \theta^{-} \end{pmatrix}. \tag{2.89}$$

Then

$$|\phi> |\theta> = \begin{pmatrix} \phi_{+}\theta_{+} \\ \phi_{+}\theta_{-} \\ \phi_{-}\theta_{+} \\ \phi_{-}\theta_{-} \end{pmatrix} \tag{2.90}$$

$$\text{and} \quad 1 \times j_i^B = \begin{pmatrix} j_i^B & 0_B \\ 0 & j_i^B \end{pmatrix} \left(\text{Each } j_i \text{ is } \frac{1}{2}\sum_i \text{ and } j_i^B \text{ acts on } |\theta> \right) \tag{2.91}$$

Then you have the more complicated looking structure

$$j_i^A \times 1 = \begin{bmatrix} (j_i^A)_{11} & 0 & (j_i^A)_{12} & 0 \\ 0 & (j_i^A)_{11} & 0 & (j_i^A)_{12} \\ (j_i^A)_{21} & 0 & (j_i^A)_{22} & 0 \\ 0 & (j_i^A)_{21} & 0 & (j_i^A)_{22} \end{bmatrix} \tag{2.92}$$

leading to:

$$J_1 = \frac{1}{2}\begin{bmatrix} 0 & 1 & 1 & 0 \\ 1 & 0 & 0 & 1 \\ 1 & 0 & 0 & 1 \\ 0 & 1 & 1 & 0 \end{bmatrix} \qquad J_2 = \frac{1}{2}\begin{bmatrix} 0 & -i & -i & 0 \\ i & 0 & 0 & -i \\ i & 0 & 0 & -i \\ 0 & i & i & 0 \end{bmatrix} \tag{2.93}$$

$$J_3 = \begin{bmatrix} 1 & & & \\ & 0 & & \\ & & 0 & \\ & & & -1 \end{bmatrix} \qquad J^2 = \begin{bmatrix} 2 & & & \\ & 1 & 1 & \\ & 1 & 1 & \\ & & & 2 \end{bmatrix}. \tag{2.94}$$

Notice that $\underline{J}^2$ is not diagonal in this $| > | >$ basis, of course.

Checks

1.

$$|1,1> = |\frac{1}{2},\frac{1}{2}> \otimes |\frac{1}{2},\frac{1}{2}> |\frac{1}{2}\frac{1}{2}> \text{implies} \begin{pmatrix}1\\0\end{pmatrix} \times \begin{pmatrix}1\\0\end{pmatrix} = \begin{pmatrix}1\\0\\0\\0\end{pmatrix} \tag{2.95}$$

Clearly $\underline{J}^2$ and J_3 have the required action.

2. The Clebsch–Gordan tables give

$$|10> = \frac{1}{\sqrt{2}}|\frac{1}{2} - \frac{1}{2}> |\frac{1}{2}\frac{1}{2}> + \frac{1}{\sqrt{2}}|\frac{1}{2}\frac{1}{2}> |\frac{1}{2} - \frac{1}{2}> . \tag{2.96}$$

Therefore represented by $\frac{1}{\sqrt{2}}\begin{pmatrix}0\\1\\1\\0\end{pmatrix}$. Again $\underline{J}^2$ and J_3 check out.

3. You may observe that

$$J_+ = \begin{bmatrix}0 & 1 & 1 & 0\\0 & 0 & 0 & 1\\0 & 0 & 0 & 1\\0 & 0 & 0 & 0\end{bmatrix} \tag{2.97}$$

$$\text{and} \quad J_-^T = \begin{bmatrix}0 & 0 & 0 & 0\\1 & 0 & 0 & 0\\1 & 0 & 0 & 0\\0 & 1 & 1 & 0\end{bmatrix} \tag{2.98}$$

have the appropriate effect.

4. The Clebsch–Gordan tables imply

$$|0,0> = \frac{1}{\sqrt{2}}|\frac{1}{2},\frac{1}{2}> \times |\frac{1}{2} - \frac{1}{2}> - \frac{1}{\sqrt{2}}|\frac{1}{2} - \frac{1}{2}> \times |\frac{1}{2}\frac{1}{2}> . \tag{2.99}$$

Therefore it is represented by $\frac{1}{\sqrt{2}}\begin{pmatrix}0\\1\\-1\\0\end{pmatrix}$. Again J^2 and J_3 check out.

Change of Basis

(Go from m_A, m_B to J, M.)

1. $< J\,M|\psi> = < J\,M|m_A m_B >< m_A m_B|\psi >$ (Use $|\,><\,| \equiv 1$.) (Note: Change of basis unitary $< J\,M|mm >< mm|J\,M >= 1$, i.e., $UU^+ = 1$.)

$$\begin{pmatrix}\psi_1\\ \psi_2\\ \psi_3\\ \psi_4\end{pmatrix} = \begin{bmatrix}(\phi\theta)_1\\ \frac{1}{\sqrt{2}}[(\phi\theta)_2+(\phi\theta)_3]\\ (\phi\theta)_4\\ \frac{1}{\sqrt{2}}[(\phi\theta)_2-(\phi\theta)_3]\end{bmatrix} = \begin{bmatrix}1&0&0&0\\ 0&\frac{1}{\sqrt{2}}&\frac{1}{\sqrt{2}}&0\\ 0&0&0&1\\ 0&\frac{1}{\sqrt{2}}&-\frac{1}{\sqrt{2}}&0\end{bmatrix}\begin{bmatrix}(\phi\theta)_1\\ (\phi\theta)_2\\ (\phi\theta)_3\\ (\phi\theta)_4\end{bmatrix}. \quad (2.100)$$

We have picked the label ordering so that

$$|1,1> \Rightarrow \begin{pmatrix}1\\0\\0\\0\end{pmatrix};\ |1,0>\Rightarrow \begin{pmatrix}0\\1\\0\\0\end{pmatrix};\ |1,-1>\Rightarrow \begin{pmatrix}0\\0\\1\\0\end{pmatrix};\ |0,0>\Rightarrow \begin{pmatrix}0\\0\\0\\1\end{pmatrix}.$$

2. $< J\,M|Op|J'M' >=$
$< J\,M|m_A m_B >< m_A m_B|Op|m'_A m'_B >< M'_A M'_B|\,J'M' >$ therefore

$$\underline{J}^2 \Rightarrow \begin{bmatrix}1&0&0&0\\ 0&\frac{1}{\sqrt{2}}&\frac{1}{\sqrt{2}}&0\\ 0&0&0&1\\ 0&\frac{1}{\sqrt{2}}&-\frac{1}{\sqrt{2}}&0\end{bmatrix}\begin{bmatrix}2&0&0&0\\ 0&1&1&0\\ 0&1&1&0\\ 0&0&0&2\end{bmatrix}\begin{bmatrix}1&0&0&0\\ 0&\frac{1}{\sqrt{2}}&\frac{1}{\sqrt{2}}&0\\ 0&0&0&1\\ 0&\frac{1}{\sqrt{2}}&-\frac{1}{\sqrt{2}}&0\end{bmatrix}$$

$$= \begin{bmatrix}2&&&\\ &2&&\\ &&2&\\ &&&0\end{bmatrix}, \quad (2.101)$$

$$\text{and } J_3^T = \begin{bmatrix}&&&\\&&&\\&&&\\&&&\end{bmatrix}\begin{bmatrix}1&&&\\ &0&&\\ &&0&\\ &&&-1\end{bmatrix}\begin{bmatrix}&&&\\&&&\\&&&\\&&&\end{bmatrix} = \begin{bmatrix}1&&&\\ &0&&\\ &&-1&\\ &&&0\end{bmatrix}. \quad (2.102)$$

Exercise

Do J_+ and J_-

$$J_+ = \sqrt{2}\begin{bmatrix} 0 & 1 & 0 & 0 \\ 0 & 0 & 1 & 0 \\ 0 & 0 & 0 & 0 \\ 0 & 0 & 0 & 0 \end{bmatrix} \qquad J_- = \sqrt{2}\begin{bmatrix} 0 & 0 & 0 & 0 \\ 1 & 0 & 0 & 0 \\ 0 & 1 & 0 & 0 \\ 0 & 0 & 0 & 0 \end{bmatrix}$$

$$J_1 = \frac{1}{\sqrt{2}}\begin{bmatrix} 0 & 1 & 0 & 0 \\ 1 & 0 & 1 & 0 \\ 0 & 1 & 0 & 0 \\ 0 & 0 & 0 & 0 \end{bmatrix} \qquad J_2 = \frac{i}{\sqrt{2}}\begin{bmatrix} 0 & -1 & 0 & 0 \\ 1 & 0 & -1 & 0 \\ 0 & 1 & 0 & 0 \\ 0 & 0 & 0 & 0 \end{bmatrix}.$$

Clearly you now have the direct sum $J^1 \bigoplus J^0$.

Note: Because $2 + 2 = 2 \times 2$ there is a standard error in this problem. One is tempted to start with the direct sum

$$\begin{pmatrix} j_i^A & 0 \\ 0 & j_i^B \end{pmatrix} \text{ acting on } \begin{pmatrix} \phi \\ \theta \end{pmatrix}$$

but this is not what is meant by the angular momentum of the system.

References

I found it impossible to find correct and appropriate references for this material and advise readers to follow the instructions in the next paragraph.

The coefficients are called Clebsch–Gordan (or vector addition) coefficients [1]. You can get tables of them. (Write to CERN Scientific Information Services, CH-1211, Geneva 23, Switzerland, and ask for a copy of the *Particle Physics Booklet* of the current academic year. It is free and there is a new one each year.)

1. *Particle Physics Booklet* p. 266.

Problems

2.1 Write out all the components of ε_{ijk}.

2.2 Show that $\delta_{ij}\Omega_j = \Omega_i$.

2.3 Show that $\delta_{ii} = 3$.

2.4 Show that $\sigma_i^\dagger = \sigma_i$.

2.5 Show that $\sigma_i \sigma_j = \delta_{ij} 1 + i\varepsilon_{ijk}\sigma_k$.

2.6 Trace $\sigma_i = 0$.

2.7 Show that $\mathrm{Det}\sigma_i = -1$.

2.8 Show that $(\sigma_i)_\alpha{}^\beta (\sigma^i)_\gamma{}^\delta 1_\alpha{}^\beta 1_\delta^\gamma = 0$.

2.9 Carefully specifying your notation, work out spin 1 in matrix form. Hint: At this time the raising or lowering of an index has no meaning. These positions are used for clarity. Start by counting the indices on the left-hand side of this equation (you should get 6). Now ask how many indices and δ's you need to match (you should get $3 \times 2 = 6$). Now ask how many indices are on the left-hand side of this equation, and recall that ε_{ijk} is antisymmetric. Note that the lower indices on the right-hand side of this equation match those on the left-hand side. Note that the numbers of the left-hand column are fixed to match these and that the upper indices start by matching those on the left-hand side but then are cycled down the left-hand column, but a pair (here m and n) are switched in the top of the right-hand column and then cycled.

2.10 Convince yourself that

$$\varepsilon_{ijk}\varepsilon^{lmn} = \begin{array}{l} \delta_i{}^l \delta_j{}^m \delta_k{}^n - \delta_i{}^l \delta_j{}^n \delta_k{}^m \\ + \delta_i{}^m \delta_j{}^n \delta_k{}^l - \delta_i{}^n \delta_j{}^m \delta_k{}^l \\ + \delta_i{}^n \delta_j{}^l \delta_k{}^m - \delta_i{}^m \delta_j{}^l \delta_k{}^n \end{array}.$$

2.11 Show that $\delta_i{}^j \delta_j{}^m = \delta_i{}^m$.

2.12 Now deduce that

$$\varepsilon_{ijk}\varepsilon^{lmk} = \delta_i{}^l \delta_j{}^n \delta_k{}^m$$

$$\text{and} \quad \varepsilon_{ijk}\varepsilon^{ljk} = 2!\delta_i{}^l$$

$$\text{and} \quad \varepsilon_{ijk}\varepsilon^{ijk} = 3!.$$

Keen students extend to N dimensions and worry about slight changes in signs of metric components.

2.13 Work out gradient, divergence, and curl where *Gradient* $\phi = \nabla_i \phi$, *Divergence* $V_i = \nabla_i V_j$, and *Curl* $V_i = \varepsilon_{ijk} \nabla_j V_k$.

2.14 Show $A = \frac{1}{\sqrt{2}} A_2 \sigma_2$ where A is antisymmetric.

2.15 Show $A_2 = \frac{1}{\sqrt{2}} Tr\, A\sigma_2$.

2.16 Show $A = \frac{1}{2}\sigma_2 Tr(A\sigma_2)$.

2.17 Show $A_\alpha^\beta = \frac{1}{2}(\sigma_2)_\alpha^\beta A_\delta^\gamma (\sigma_2)_\gamma^\delta$.

2.18 Taking a complete set of observables as $\{j_z^A, (\underline{j}^A)^2, J_z^B, (\underline{j}^B)^2\}$ show that the states in this basis have the outer product form.

2.19 Show that when $|j^A j^B m^A m^B> \equiv |j^A m^A> |j^B m^B>$ then the action of the operators is $j_i^A \rightarrow j_A^i \times 1$ and $j_i^B \rightarrow 1 \times j_i^B$.

2.20 When the total angular momentum is defined by $J_i = j_i^A + j_i^B$, show that the commutator $[J_i, J_j]$ has the usual form.

2.21 Drawing the straight line figure for M running from $(-)(j_A + j_B)$ through $(-)\, j_A$ and $(-)\, j_B$ and zero, and j_B and j_A to $(j_A + j_B)$ show that the number of times you get $j_A + j_B$ is just once for M, but twice for $j_A + j_B - 1$ and so forth until you reach $M = j_A - j_B$.

2.22 Show that the number of times you get M_n is given by $n(M) = j_A + j_B + 1 - |M|$, if $j_A + j_B \geq M \geq |j_A - j_B|$.

2.23 Show that now you finally reach $(2j_B + 1)$ when you have used all $(2j_B + 1)$ values of m_B and moving further left you find a sufficiently negative m_B to make $m_A = j_A$.

2.24 Show that you now have an angular momentum of the form $n(M) = 2j_B + 1$ if $|M| \leq |j_A - j_B|$.

2.25 Finally show that $N(J) = 1$ for $J = j_A + j_B, j_A + j_B - 1, \ldots |j_A - j_B|$. Hint: The changes of $n(m)$ are what really count.

2.26 Consider the case

$$1 \otimes \frac{1}{2} = \frac{3}{2} \oplus \frac{1}{2}.$$

Show that the stretched state is

$$|\frac{3}{2}, \frac{3}{2} >= |1, 1 > |\frac{1}{2}\frac{1}{2} > .$$

2.27 Apply the operator J_- to show that

$$|\frac{3}{2}, \frac{1}{2} >= \sqrt{\frac{2}{3}}|1, 0 > |\frac{1}{2}, \frac{1}{2} + \sqrt{\frac{1}{3}}\, |1, 1 > |\frac{1}{2}, -\frac{1}{2} > .$$

2.28 Find $|\frac{1}{2}, \pm\frac{1}{2} >$ by writing down a linear form with two parameters and applying normalization and orthogonality. You may assume the Condon and Shortley phase conventions that if $j_A \geq j_B$ then $\{< j_A j_A | j_B, J - j_A|\}|J\, J >$ is real and > 0.

2.29 Finally show that

$$|1, 0 > |\frac{1}{2}, -\frac{1}{2} >= \frac{1}{\sqrt{3}}|\frac{1}{2}, -\frac{1}{2} > +\sqrt{\frac{2}{3}}|\frac{3}{2}, -\frac{1}{2} > .$$

2.30 Work out $\frac{1}{2} \otimes \frac{1}{2}$ in full, construct the table, then check against the CERN booklet mentioned in the References section.

2.31 Show $A \times B \neq B \times A$ in general.

2.32 Show $(A + B) \times C = AC + BC$.

2.33 If A and A' are both $(m \times n)$ and B and B' are both $(n \times n)$, show that $(A \times B)(A' \times B') = (AA') + (BB')$.

2.34 Show $Tr(A \times B) = Tr\, A.Tr\, B$.

2.35 If $A^\dagger = A^{-1}$ and $B^\dagger = B^{-1}$, then show $(A \times B)^\dagger = (A \times B)^{-1}$.

2.36 Work through in detail $\frac{1}{2} \otimes \frac{1}{2} = 1 \oplus 0$ in matrix form to check that it gives what you expect.

3

Tensors and Tensor Operators

Think about a collection of points in a space. Usually the ordinary points of our three-dimensional space are what we shall need, but sometimes the "points" might be the states of a quantum mechanical Hilbert space. (You might have to deal with a manifold, hyperspace, variety, etc.) Give the points labels so that you can find them again. Say we use x^i where i takes values in the appropriate range, for example, $i = 1, 2, 3$. Then move the points of the space (I take the active viewpoint), and the point that had labels x^i now has $x^{i'}$ as labels. For the physics we have in mind, we assume that the functions give the new coordinates $x^{i'}$ in terms of the old x^i.

$$x'^r = f^r(x^1, x^2, \ldots, x^N) \qquad (r = 1, \ldots, N) \tag{3.1}$$

are continuous and differentiable (real, if the coordinates are real), and the functional determinant (Jacobian)

$$\begin{vmatrix} \frac{\partial x''}{\partial x'} & \cdots & \frac{\partial x''}{\partial x^N} \\ \vdots & \ddots & \vdots \\ \frac{\partial x'^N}{\partial x'} & \cdots & \frac{\partial x'^N}{\partial x^N} \end{vmatrix} \tag{3.2}$$

never vanishes. (Of course, you put up with singularities such as that at the origin in the change between Cartesian and spherical polar coordinates.) In principle you can then solve to get

$$x^r = g^r(x'', x'^2, \ldots, x'^N) \qquad (r = 1, 2, \ldots, N) \tag{3.3}$$

and this is called the implicit function theorem.

Now what are the objects that are of interest to a physicist? Those things that have simple properties under the transformations. We will list some of the more important ones.

Scalars

Often called *invariants*, scalars are numbers, or more importantly results of measurements, that are numerically the same for the new system measured with the old apparatus.

DOI: 10.1201/9781439895207-3

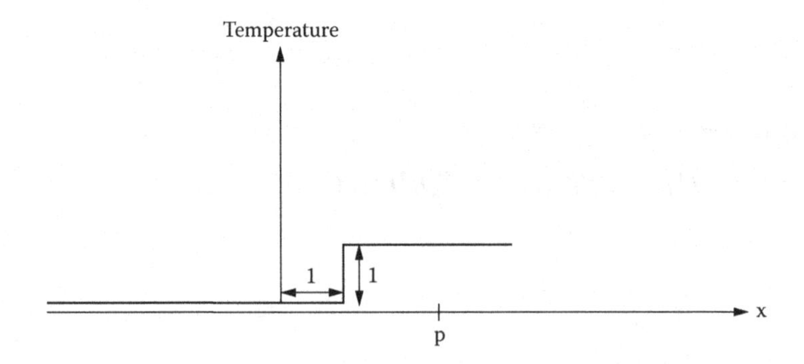

FIGURE 3.1
The temperature plotted againt position.

Scalar Fields

Sometimes called *scalar functions*, scalar fields are a set of numbers (one at each point) that "go round with the system." Examples are physical functions of space on a body, for example, temperature. We describe this by a single (no index) function of the coordinates, $\phi(x', \ldots, x^N)$, but this is not a definite (fixed) mathematical function even though we do not bother to change the symbol ϕ. Suppose, for example, you heat a bar uniformly to 1 degree to the right of 1 meter from the origin, so the temperature function is given by

$$\phi(x) = \theta(x-1). \tag{3.4}$$

(See Figure 3.1.)

Now transform the system a distance d to the right. For each point $x \to x' = x + d$. The plot of temperature now looks like

$$\phi'(x') = \phi'(x+d) \equiv \phi(x) = \theta(x-1) \tag{3.5}$$

of course. As a function we now have

$$\phi'(x) = \phi(x-d) = \theta(x-1-d), \tag{3.6}$$

which you can see in Figure 3.2. In general,

$$\phi(x) \to \phi'(x) \equiv \phi(x) \tag{3.7}$$

and if

$$x \to x' = Tx, \quad \text{then} \quad \phi'(x) = \phi(T^{-1}x). \tag{3.8}$$

(The physical function is invariant; the mathematical one changes.)

Invariant Functions

Invariant functions (e.g., $r^2 = x^2+y^2+z^2$ under [3] rotations) have $f(x') = f(x)$ for all allowable points x' and x. This can be thought of as a *form invariant*

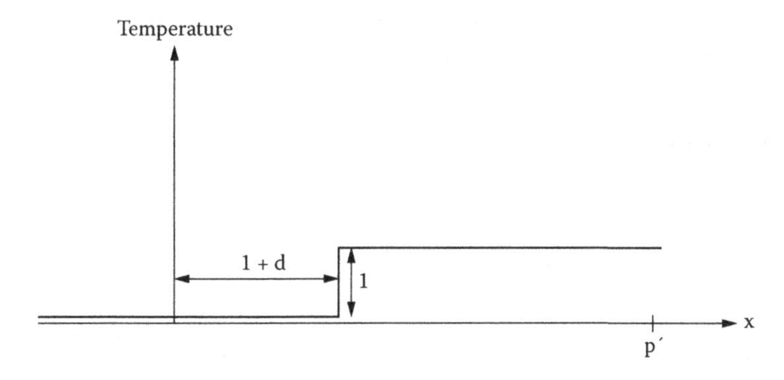

FIGURE 3.2
The temperature plotted against position translated.

scalar field. Such objects might be the wave functions of states that are invariant under rotations.

Contravariant Vectors ($t \to$ Index at Top)

Two neighboring points define a vector $\overrightarrow{PQ}$ at P, which has coordinates dx^r (see Figure 3.3).

If we transform, then

$$dx'r = \frac{\partial x'^r}{\partial x^s} dx^s \tag{3.9}$$

where we notice that $\frac{\partial x'^r}{\partial x^s}$ is fixed (for each given point P) by the transformation; it does not depend on Q, that is, it does not depend upon the vector $\underline{dx}$. This transformation is linear and homogeneous (affine) in the dx^r. More generally, we define contravariant vectors by sets of components B^i (associated with a point P) that transform like

$$B^i \to B^{i'} = B^j \frac{\partial x'^i}{\partial x^j} \overset{df}{=} B^j (D^{-1})^i_j \tag{3.10}$$

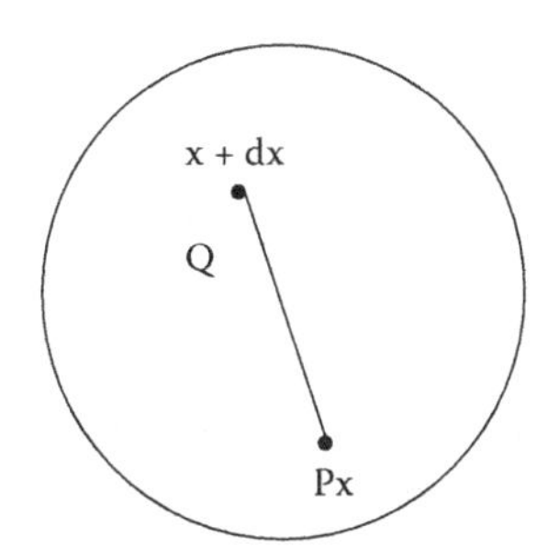

FIGURE 3.3
Two neighboring points define a vector PQ at $\underline{P}$.

where the B^j and the $(D^{-1})^i_j$ are evaluated at P if it matters. This last (ludicrous) definition makes sense when you go on to introduce covariant vectors.

Covariant Vectors (Co = Goes Below)

Covariant vectors have lower indices and a rule:

$$A_i \rightarrow A'_i = \frac{\partial x^j}{\partial x'^i} A_j = D^j_i A_j \tag{3.11}$$

where the last step follows the chain rule of differentiation:

$$\delta^j_i = \frac{\partial x^j}{\partial x'^i} = \frac{\partial x'^k}{\partial x^i}\frac{\partial x^j}{\partial x'^k}, \quad \text{or} \quad 1^j_i = (D^{-1})^k_i (D)^j_k. \tag{3.12}$$

An example (usually the prototype) of a covariant vector is provided by the gradient of a scalar

$$\frac{\partial \phi}{\partial x^r} = \partial_r \phi \tag{3.13}$$

since

$$\frac{\partial \phi}{\partial x^i} \rightarrow \frac{\partial}{\partial x'^i}\phi'(x') = \frac{\partial}{\partial x'^i}\phi(x) = \frac{\partial \phi(x)}{\partial x^j}\frac{\partial x^j}{\partial x'^i}. \tag{3.14}$$

You usually deduce this from the remark that the difference between two scalars at nearby points

$$\phi(x + dx) - \phi(x) \tag{3.15}$$

had clearly better be invariant. Thus in the limit $\frac{\partial \phi}{\partial x^r} dx^r$ has to be invariant, and hence

$$\partial_r \phi \rightarrow D^s_r \partial_s \phi \tag{3.16}$$

has to follow.

Notes

1. In general the x^r themselves are not components of a vector at all.
2. In general D depends on the point $\underset{\sim}{x}$, often in a horrible nonlinear way. The presentation given is now viewed as old fashioned by mathematicians but it serves our purpose very well.

Tensors

The previous ideas extend to tensors of any rank (number of indices) of mixed contravariant and covariant types:

$$T_{ab\ldots}^{rst\ldots} \to T'_{ab\ldots}{}^{rst} = D_a^{a'} D_b^{b'} T_{a'b'\ldots}^{r's't'} (D^{-1})_{r'}^{r} (D^{-1})_{s'}^{s} (D^{-1})_{t'}^{t}. \tag{3.17}$$

Notes and Properties

1. Scalars are zero index tensors, and contra(co)variant vectors are single index tensors.
2. Because the transformation laws are linear in the tensor components, you can add or subtract constant multiples of identical tensor types.

$$C_{de}^{abc} = A_{de}^{abc} + 7B_{de}^{abc}. \tag{3.18}$$

3. You can form outer products, which get the appropriate transformations,

$$V_r W_s = T_{rs}. \tag{3.19}$$

4. Just as $(\partial_r \phi) dx^r$ was a scalar, there is an extension to inner products by contraction of contravariant against covariant indices,

$$T_{rs} A^s = B_r. \tag{3.20}$$

5. Symmetry holds after transformation,

$$T^{\mu\nu} = \pm T^{\nu\mu}. \tag{3.21}$$

6. δ_i^j is a mixed tensor.
7. Theorem: Given some objects $S_{\ldots}^{\cdots}$, and that $S_{\ldots}^{\cdots} T_{\ldots}^{\cdots}$ is invariant for all $T_{\ldots}^{\cdots}$, then $S_{\ldots}^{\cdots}$ form a tensor.

   Proof: Suppose that $S \to \overline{S}$ instead of S' under a transformation. Both $S'T'$ and $\overline{S}T'$ are equal to ST. Hence, $(\overline{S} - S')T' = 0$, but T' are arbitrary.
8. Reduction of tensors is performed by symmetrization (or antisymmetrization) of all index pairs of the same type and extraction of all δ_i^j pieces by contraction, until an irreducible tensor is reached. (If there is an ε_{ijk} tensor [see later] this must be taken out too. Also [shortly]

we shall learn how δ_{ij} and δ^{ij} exist for rotations; then these must be pulled out. In general, any invariant [numerical] tensor must be used if it exists.)

$$\text{(a)}\quad T^{ij} = S^{ij} + A^{ij} \text{ where} \tag{3.22}$$

$$S^{ij} = \frac{1}{2}(T^{ij} + T^{ji}) \text{ and} \tag{3.23}$$

$$A^{ij} = \frac{1}{2}(T^{ij} - T^{ji}). \tag{3.24}$$

$$\text{(b)}\quad M_i^j = \overline{M}_i^j + \frac{1}{3}\delta_i^j M_k^k \tag{3.25}$$

$$\text{has no trace, and } X = M_k^k. \tag{3.26}$$

Notice that, in (a), A^{ij} has three components and can be related to a covariant vector by

$$V_i = \frac{1}{2}\varepsilon_{ijk} A^{jk} \tag{3.27}$$

$$A^{ij} = \varepsilon_{ijk} V_k \tag{3.28}$$

if you like. In all examples, practice and ingenuity are important. Try some.

9. Repeated transformations (transitivity). Shift from $\underline{x}$ to $\underline{x}'$, then $\underline{x}'$ to $\underline{x}''$.

$$V^r \to V'^r = \frac{\partial x'^r}{\partial x^s} V^s \tag{3.29}$$

$$\text{then}\quad V'^r \to V''^r = \frac{\partial x''^r}{\partial x'^p} V'^p \tag{3.30}$$

$$= \frac{\partial x''^r}{\partial x'^p}\frac{\partial x'^p}{\partial x^s} V^s \tag{3.31}$$

$$= \frac{\partial x''^r}{\partial x^s} V^s. \tag{3.32}$$

(This obviously always works!) Why is this useful? Now that we have established transitivity, since the transformation laws are linear and homogeneous, an equation true before transformation is also true afterward. (From a passive viewpoint, an equation true in one frame is true in all others.) So such statements are about the physics, not just the way in which the observer sees it.

Rotations

At this point I feel that you have enough of the general definitions to work on yourself whenever you need, so I shall move on to rotations in Euclidean space. Then the relationship between the x'^i and the x^i is linear, real, and preserves distances. (We also stick to right-handed axes.) The linearity means that we write

$$x'^i = x^j (M^{-1})^i_j \tag{3.33}$$

$$x^i = x'^j (M)^i_j \tag{3.34}$$

with M a matrix of numbers (i.e., independent of x). Obviously,

$$\frac{\partial x^j}{\partial x'^i} = M^j_i \tag{3.35}$$

so that D^j_i is identical to M^j_i in this case. The x^i are the components of a contravariant vector for rotations. The reality of the transformations means that D has real elements. (You can also show that sticking to right-handed coordinates means that the determinant of D can be fixed to be unitary.) The length (Euclidean) of a vector is defined by the sum of squares of its components, of course, so we have

$$x^i x^i = x'^i x'^i = x^j (M^{-1})^i_j x^k (M^{-1})^i_k . \tag{3.36}$$

In matrix language

$$MM^T = 1 \tag{3.37}$$

and the matrix is orthogonal. Now that is handy. In matrix language you had

$$A \to DA \tag{3.38}$$

$$B \to (D^{-1})^T B \tag{3.39}$$

but $D = M$, so that

$$A \to MA \tag{3.40}$$

$$B \to MB \tag{3.41}$$

and for rotations covariant and contravariant are equivalent. We shall just speak of vectors. You can put the vector or tensor indices up or down at will. The numerically invariant tensors δ^j_i and ε_{ijk} are now available as δ_{ij}, δ^{ij}, ε^j_{ik}, and so forth.

Now we return to the study of interesting objects with simple transformation laws.

Vector Fields

Vector fields are physical functions of space (and time) that are vector valued. A set of (three) numbers at every point of space are components of a vector at that point and "go round with the system" (e.g., the velocities at all points of a fluid at one time). To make this more precise, we introduce more notation for rotations in our [3] space. Take a set of axes (orthonormal) $\underline{e}_i$; specify a rotation by a set of (orthonormal) vectors $\underline{n}_i$, which you get by rotating vectors originally coincident with the $\underline{e}_i$. (The $\underline{e}_i$ are fixed for all time.)

(Note: The i on $\underline{e}_i$ or $\underline{n}_i$ is a label.)

Then

$$\underline{n}_i = R(\underline{e}_i) = \underline{e}_j R_{ji} \tag{3.42}$$

Clearly

$$R_{ij} = \underline{e}_i \cdot \underline{n}_j; \tag{3.43}$$

and if you wish to specify the rotation in detail you would probably give the axis of rotation and the right-hand-thread rotation around it—$R(\underline{u}, \theta)$—or you might use Euler angles.

Let us connect this with our previous ideas about vectors. The components of vectors relate to the labels on the fixed basis vectors by

$$\underline{V} = V_i \underline{e}_i \tag{3.44}$$

$$\underline{V} = V_i \cdot \underline{e}_i. \tag{3.45}$$

Now, a vector "goes round with the system" so that

$$V_i' \underline{e}_i \equiv \underline{V}' = R(\underline{V}) = V_j \underline{n}_j = V_j \underline{e}_i R_{ij} = (R_{ij} V_j) \underline{e}_i, \tag{3.46}$$

therefore

$$V_i' = R_{ij} V_j. \tag{3.47}$$

Good! $R_{ij} = M_{ij} = D_{ij}$. And, of course, all of this extends to tensors. (We are not deliberately awkward in having three "matrices" (R, M, D)—when you wish to read on then it is important to know that they are logically distinct but coincide in the present case.)

Now to return to vector fields. The precise form of the transformation is given by:

$$\underline{V}(\underline{x}) = V_i(\underline{x}) \underline{e}_i \tag{3.48}$$

$$V_i(x) = \underline{V}(\underline{x}) \cdot \underline{e}_i \tag{3.49}$$

so that

$$\underline{e}_i V_i'(\underline{x}) = \underline{V}'(\underline{x}) \equiv R \underline{V}(R^{-1} \underline{x}) \tag{3.50}$$

as above.

(In words: The new vector at $\underline{x}$ is the rotation of the old vector at the point that moves to $\underline{x}$.) Therefore

$$V_i'(\underline{x}) = R_{ij} V_j(R^{-1}\underline{x}) \tag{3.51}$$

$$\{\text{Scalar field } \psi'(\underline{x}) = \psi(R^{-1}x)\}. \tag{3.52}$$

(In general you distinguish covariance and contravariance.)

It is probably worth reiterating that although the physical function is unchanged ($\psi'(x') = \psi(x)$) or merely has its components mixed ($V_i'(x') = R_{ij} V_j(x)$) we speak of the change (mathematical) of the objects as

$$\psi \to \psi' \quad \text{with} \quad \psi'(x) = \psi(R^{-1}x) \tag{3.53}$$

$$V_i \to V_i' \quad \text{with} \quad V_i'(x) = R_{ij} V_j(R^{-1}x) \tag{3.54}$$

and you have to keep this in mind. (But physically $\psi(x) \to \psi'(x') \equiv \psi(x)$.)

Tensor fields are obvious generalizations of vector fields. Now for a new class of objects.

Tensor Operators

(This is still classical, not quantum.)

Scalar Operator

Not affected by rotation. If it gives a measurement, then the result is a scalar. If it is composed of the components of tensors, then it is so cunningly constructed that it "undoes" their transformations. Compare this with invariant function and contrast this with scalar field.

Vector Operator

These objects (N in N dimensions), which are called the components, go round with the system like vectors would (not like the components of vectors). Remember that

$$\underline{n}_i = \underline{e}_j R_{ji} \tag{3.55}$$

for vectors originally coincident with the $\underline{e}_i$. So we have

$$K_i' = K_j R_{ji} = (R^{-1})_{ij} K_j. \tag{3.56}$$

As an example, recall how our prototype of a covariant vector was

$$\frac{\partial}{\partial x^i}\phi(x)$$

where $\phi(x)$ was a scalar field. We can think of a (fixed) vector operator $\underline{\nabla}$ such that

$$\underline{\nabla}\phi = \underline{e}_i\frac{\partial\phi}{\partial x^i} \qquad \left\{ \underline{\text{or}}\ (\underline{\nabla}\phi)_i = \frac{\partial\phi}{\partial x_i} \right\}.$$

Now we already know the effect here is to write down the rate of changes, and we can think of writing

$$\underline{\nabla} = \underline{e}^i\,\nabla_i = \underline{n}^i\,\nabla_i' \text{ (note order for safety)} \tag{3.57}$$

so that you envisage a set of three operators that change to compensate the $\underline{e} \to \underline{n}$. Clearly then

$$\underline{e}^i\,\nabla_i = \underline{n}^i\,\nabla_i' \tag{3.58}$$

$$= \underline{e}^j\,D^i_j\nabla_i', \tag{3.59}$$

therefore

$$\nabla_i = D_i^j\,\nabla_j', \tag{3.60}$$

therefore

$$\nabla' = (D^{-1})_i^j\,\nabla_j \qquad \text{(inverse).} \tag{3.61}$$

This concept extends to tensor operators of all kinds. Examples abound. (In quantum mechanics the usual r_i, L_i, p_i are all vector operators.)

Notes

1. In general there is no relationship between collections of objects forming tensors and those forming tensor operators.
2. Reduction of tensor operators to irreducible (or spherical) tensor operators follows exactly as it did for tensors.
3. If you compare

$$K'^i = K^j\,D^i_j \qquad \text{and} \tag{3.62}$$

$$K_i' = (D^{-1})_i^j\,K_j \tag{3.63}$$

with Equation (3.52), that is,

$$B'^{i} = B^{j}(D^{-1})^{j}_{i} \qquad \text{and} \tag{3.64}$$

$$A'_{i} = D^{j}_{i} A_{j}, \tag{3.65}$$

this is said to be cogredient transformation by some authors. Warning: Someone taking a passive point of view can "send his axes" through $(-\theta)$ to "agree" with us—then the transformation laws for K_i and A_i coincide. This looks neat but hides the problem.

Connection with Quantum Mechanics

The states of systems get "moved around" by transformations—you can think of the states as points. The Dirac description is in terms of the state vectors $|\psi>$ (in Hilbert space), and the components (which might be $\psi_i(\underline{x}) =< \underline{e}_i, \underline{x}|\psi >$, say) are the corresponding wave functions that you can think of as coordinates if you like.

Transformations are represented by operators (quantum mechanical this time) that act on the states to give new ones. Although norms (scalar products) are preserved, you must remember that complex conjugation is involved:

$$\text{If } |\psi \rightarrow |\psi' > = U|\psi >, \tag{3.66}$$

$$\text{then} \quad < \psi'|\psi' =< \psi|\psi \Rightarrow \mathbf{U}^{+}\mathbf{U} = \mathbf{1}. \tag{3.67}$$

So the operators representing transformations in this space are unitary.

Observables

(Represented by Hermitian operators). All dynamic variables have to transform with the states. If you have

$$A|\psi > = |\phi > \tag{3.68}$$

$$A'|\psi' > = |\phi' > \tag{3.69}$$

for arbitrary states, then

$$A'U|\psi > = U|\phi >= UA|\psi >, \tag{3.70}$$

so that

$$\mathbf{A}' = \mathbf{U}\mathbf{A}\mathbf{U}^{-1}. \tag{3.71}$$

In particular, if the observable gives a measurement then it must transform so that the eigenvalue for the rotated state is the same as it was (for the original observable) on the old state:

$$A|a_n >= a_n|a_n > \quad (3.72)$$

$$UA|a_n >= a_n U|a_n > \quad (3.73)$$

$$(UAU^{-1})U|A_n >= a_n U|a_n >, \quad (3.74)$$

therefore

$$\mathbf{A}' = \mathbf{UAU}^{-1}. \quad (3.75)$$

Rotations

(Think about ordinary ones in three-dimensional space.) Corresponding to each rotation, R, in ordinary three-dimensional space there is a unitary operator, $U(R)$, which transforms the states. All the "nice" properties are preserved correctly, so there is no need for most of us to worry about the technical side of things most of the time.

$$U(1) = 1 \quad (3.76)$$

$$U(R^{-1}) = U^{-1}(R) \quad (3.77)$$

$$U(R_2R_1) = U(R_2)U(R_1) \quad (3.78)$$

with the usual convention that the one written to the right is applied first in each case.

Scalar Fields

Scalar fields often appear as quantum mechanical wave functions. It is important to notice that we regard the reference bras $< \underline{x}, \underline{e}_i|$ as fixed when we envisage a rotation, but we can then think of the action of the operators on them. Thus under a rotation R we get:

$$|\psi > \rightarrow |\psi' >= U(R)|\psi >, \quad (3.79)$$

$$< x| \rightarrow < x|, \quad (3.80)$$

and so

$$\psi(x) =< x|\psi > \rightarrow \psi'(x) =< x|\psi' > =< x|U|\psi > \quad (3.81)$$

where we then go on to say:

$$U|x > = |Rx > \quad (3.82)$$

but

$$UU^{\dagger} = 1, \quad \text{that is,} \quad U^{\dagger} = U^{-1} \tag{3.83}$$

therefore

$$U^{\dagger}|x> = |R^{-1}x>, \tag{3.84}$$

so that

$$< R^{-1}x| =< x|U, \tag{3.85}$$

and hence

$$\psi'(x) =< x|U|\psi >=< R^{-1}x|\psi >= \psi(R^{-1}x), \tag{3.86}$$

to establish the connection with our scalar field as defined earlier.

Vector Fields

Vector fields often appear in a similar way. If we are describing a system with angular momentum, then the wave function will have three components that form the components of a vector field at each point. (This could be a hydrogen atom in a state with labels $n = 2, l = 1, m = (+1, 0, -1)$.) We would have

$$< x, \underline{e}_i|\psi > = \psi_i(x) \tag{3.87}$$

where we have taken our usual fixed axes in a Cartesian basis. Then, under a rotation R, we get

$$|\psi > \to |\psi' >= U(R)|\psi > \tag{3.88}$$

$$< x, \underline{e}_i| \to< x, \underline{e}_i| \tag{3.89}$$

$$\psi_i(x) =< x, \underline{e}_i|\psi > \to< x, \underline{e}_i|\psi' >= \psi_i'(x). \tag{3.90}$$

To go on from here you have to be quite specific in handling the reference states (vectors)—let's do this in full detail just for once:

$$U(R^{-1})|\underline{x}, \underline{e}_i > = |R^{-1}(\underline{x}), R^{-1}\underline{e}_i > \tag{3.91}$$

$$= |R^{-1}(\underline{x}), \underline{e}_j R^{-1}_{ji} > \tag{3.92}$$

$$= |R^{-1}(\underline{x}), R_{ij}\underline{e}_j > \tag{3.93}$$

$$= R_{ij}|R^{-1}(\underline{x}), \underline{e}_j > . \tag{3.94}$$

That is

$$< \underline{x}, \underline{e}_i|U(R) = R_{ij} < R^{-1}(\underline{x}), \underline{e}_j| >, \tag{3.95}$$

since R is real and does not get transposed here.

Then

$$\psi_i'(x) =< \underline{x}, \underline{e}_i|\psi' > \tag{3.96}$$

$$=< \underline{x}, \underline{e}_i|U|\psi > \tag{3.97}$$

$$= R_{ij} < R^{-1}\underline{x}, \underline{e}_j|\psi, \tag{3.98}$$

therefore

$$\psi_i'(x) = R_{ij}\psi_j(R^{-1}x), \tag{3.99}$$

and we have made the connection with what we called the transformation law for vector fields earlier.

The extension to see how tensor fields occur as wave functions of states with high angular momentum should now be obvious.

A form invariant scalar field (invariant function) can appear in the form of an invariant wave function. If we have a state (such as the ground state of hydrogen with a wave function $\psi(\underline{x})$ $\exp(\frac{-r}{a})$) that is invariant under rotations,

$$|\psi >\rightarrow |\psi' > = U|\psi >\equiv |\psi >, \tag{3.100}$$

then

$$\psi(x) =< x|\psi >\rightarrow< x|\psi' = \psi'(x) >, \tag{3.101}$$

and now therefore

$$\psi'(x) = \psi(x) \tag{3.102}$$

and we retrieve form invariance.

Scalar operators (invariant operators) often appear as those which give measurements which are scalars. These are fixed operators $A' \equiv A$, and since a general state transform as $|\psi >\rightarrow U|\psi >$, we deduce that

$$[A, U] = 0. \tag{3.103}$$

The generators of the transformations (J_i for rotations) all commute with scalar operators. (A good example would be the Hamiltonian operator of a rotationally invariant system [ground state hydrogen atom]. This measures the energy that is unchanged by rotation.)

Obviously you try to do measurements associated with scalar operators, so that, although you never move your fixed observing apparatus, the operator remains appropriate since its change (as a dynamical variable) would have been no change at all in this case.

Tensor operators arise when you try to do rotations of a system and insist that the expectation value of some fixed operators (associated with apparatus in your fixed reference frame) form the components of a vector (or more generally, a tensor). That is, you have:

$$< \psi|P_i|\psi > \rightarrow R_{ij} < \psi|P_j|\psi >, \tag{3.104}$$

but you know that $|\psi \rightarrow U(R)|\psi$, so that

$$< \psi|P_i|\psi > \rightarrow < \psi|U^{-1}P_iU|\psi > \tag{3.105}$$

and hence

$$U^{-1}P_iU = R_{ij}P_j \tag{3.106}$$

$$\text{or} \quad UP_iU^{-1} = R_{ij}^{-1}P_j = P_jR_{ji}. \tag{3.107}$$

If you have observables, or dynamical variables (these are quantum mechanical operators), which coincide with these originally, then they form the components of a vector operator according to our earlier definitions. (Their expectation values between fixed reference states also form a vector operator.) Examples include position momentum and angular momentum operators. The extension to tensor operators of arbitrary rank is now straightforward.

Note: You can see how the game is starting to get complicated; you can dream up all sorts of weird objects. But the ones I have indicated are the most important ones for physics. One last warning: If you do quantum field theory then the whole game gets yet one more level of complication. But our definitions and so forth are fine if you work with care.

Specification of Rotations

There are many ways to specify a rotation, but we will stick to the crude idea of defining an axis by a unit vector $\underline{n}$, and then specifying an angle of rotation θ about that axis in the right-hand thread sense [1,2]. Note: (1) Do not confuse this $\underline{n}$ with the moving axes. (2) Read about Euler angles.

For example, if we have $V_i \rightarrow R_{ij}V_j$ by an angle θ around the z-axis, then because $R = M = D$, and $R^T = R^{-1}$ we know

$$x'^r = x^s(R^{-1})^r_s \qquad \text{or} \qquad x'_r = R_{rs}x_s \tag{3.108}$$

$$[x'] = [x]R^{-1} \qquad \text{or} \qquad [x'] = (R^{-1})^T[x] = R[x] \tag{3.109}$$

becomes

$$x' = x\cos\theta - y\sin\theta \tag{3.110}$$

$$y' = y\cos\theta + x\sin\theta \tag{3.111}$$

$$z' = z. \tag{3.112}$$

Hence, we write

$$R(\underline{e}_3, \theta) = \begin{pmatrix} \cos\theta & -\sin\theta & 0 \\ \sin\theta & \cos\theta & 0 \\ 0 & 0 & 1 \end{pmatrix} \tag{3.113}$$

and

$$\left[\begin{pmatrix} V \\ \\ \\ \end{pmatrix}\right] \rightarrow \left[\begin{pmatrix} & & \\ & R & \\ & & \end{pmatrix}\right]\left[\begin{pmatrix} V \\ \\ \\ \end{pmatrix}\right] \doteq \left[\begin{pmatrix} 1 & -\theta & 0 \\ \theta & 1 & 0 \\ 0 & 0 & 1 \end{pmatrix}\right]\left[\begin{pmatrix} V \\ \\ \\ \end{pmatrix}\right] \tag{3.114}$$

$$\doteq \left[\begin{pmatrix} V \\ \\ \\ \end{pmatrix}\right] + \theta \left[\begin{pmatrix} 0 & -1 & 0 \\ 1 & 0 & 0 \\ 0 & 0 & 0 \end{pmatrix}\right]\left[\begin{pmatrix} V \\ \\ \\ \end{pmatrix}\right] \tag{3.115}$$

working with infinitessimals. We can write this generally as

$$V_i \rightarrow V_i' = V_i + \theta \varepsilon_{ijk} n_j V_k + O(\theta^2) \tag{3.116}$$

$$= V_i + \varepsilon_{ijk} \theta_j V_k \tag{3.117}$$

that is,

$$\underline{V}' = \underline{V} + \theta \underline{n} \times \underline{V} + O(\theta^2). \tag{3.118}$$

Transformation of Scalar Wave Functions

Suppose we have a scalar wave function so that when we transform the system the mathematical change in the function is given by

$$\psi(x) \rightarrow \psi'(x) = \psi(R^{-1}x). \tag{3.119}$$

For our small angle rotation about the z-axis we get

$$\psi'(x) = \psi(x + \theta y, y - \theta x, z) + \ldots \text{(Just } R^{-1}(\theta) = R(-\theta)\text{)} \tag{3.120}$$

$$= \psi(x, y, z) + \theta\left(y\frac{\partial\psi}{\partial x} - x\frac{\partial\psi}{\partial y}\right) + \ldots \text{(Taylor expansion)} \tag{3.121}$$

$$= \psi(x, y, z) - \frac{i\theta}{\hbar} L_z \psi(x, y, z) + \ldots \tag{3.122}$$

where L_z is the usual orbital angular momentum operator.

We write

$$\psi(x) \to R_z(\theta)\psi(x) \tag{3.123}$$

with

$$R_z(\theta) = 1 - \frac{i\theta}{\hbar}L_z + \dots, \tag{3.124}$$

and this extends to

$$R(\underline{n}, \theta) = 1 - \frac{i\theta}{\hbar}\underline{n} \cdot \underline{L} + \dots \tag{3.125}$$

if we rotate about an axis specified by $\underline{n}$.

We define

$$R(\underline{n}, \theta) = 1 - \frac{i\theta}{\hbar}\underline{n} \cdot \underline{J} + \dots \tag{3.126}$$

when the system has no classical analogue, that is, for odd-half-integer spins.

Note: At the moment we do not know that all of this is consistent. The result of two rotations around different axes has to have the appropriate effect as a rotation. We must check that this ties up with the usual commutation relations we assigned to components of angular momentum. We shall come to this shortly.

Finite Angle Rotations

We build up $R(\underline{n}, \theta)$—now for finite θ—by repeated application of small rotations around the same axis $\underline{n}$.

$$R(\underline{n}, \theta + d\theta) = R(\underline{n}, d\theta)R(\underline{n}, \theta) \tag{3.127}$$

$$= \left(1 - \frac{i}{\hbar}\underline{n} \cdot \underline{J}\, d\theta\right) R(\underline{n}, \theta) \tag{3.128}$$

therefore

$$\frac{d}{d\theta} R(\underline{n}, \theta) = -\frac{i}{\hbar}\underline{n} \cdot \underline{J}\, R(\underline{n}, \theta), \tag{3.129}$$

and even we can solve this equation to get

$$R(\underline{n}, \theta) = \exp\left(-\frac{i\theta}{\hbar}\right)\underline{n} \cdot \underline{J} \tag{3.130}$$

since

$$R(\underline{n}, 0) = 1. \tag{3.131}$$

Consistency with the Angular Momentum Commutation Rules

Consider our rotation $U(R)$ specified by $R(\underline{n}, \theta)$, and think about the components K_i of some vector operator. We know that

$$UK_jU^{-1} = K_kR_{kj} = (R^{-1})_{jk}K_k, \tag{3.132}$$

$$\exp\left(-\frac{i\theta}{\hbar}\underline{n}\cdot\underline{J}\right)K_j\exp\left(\frac{i\theta}{\hbar}\underline{n}\cdot\underline{J}\right) = (R^{-1})_{jk}K_k, \tag{3.133}$$

so that expanding

$$\left(1-\frac{i\theta}{\hbar}\underline{n}\cdot\underline{J}+\ldots\right)K_j\left(1+\frac{i\theta}{\hbar}\underline{n}\cdot\underline{J}+\ldots\right) = K_j - \varepsilon_{jik}\theta_iK_k + \ldots \tag{3.134}$$

where the last step follows from the law we deduced for V_i with $\theta \to (-)\theta$. Thus

$$K_j - \frac{i\theta}{\hbar}n_i[J_i, K_j] + \ldots = K_j - \varepsilon_{jik}\theta n_iK_k + \ldots \tag{3.135}$$

and we deduce

$$[J_i, K_j] = i\hbar\varepsilon_{ijk}K_k \tag{3.136}$$

since $\underline{n}$ was arbitrary.

Now we can put it all together. The components of the angular momentum are themselves a vector operator, since the expectation values of them must transform like the classical angular momentum vector. Putting $\underline{K} \to \underline{J}$ we see

$$[J_i, J_j] = i\hbar\varepsilon_{ijk}J_k \tag{3.137}$$

and we have retrieved our previous commutation relations.

Rotation of Spinor Wave Function

The spin $\frac{1}{2}$ wave function with two components uses the Pauli matrices as

$$S_i = \frac{\hbar}{2}\sigma_i \tag{3.138}$$

to describe the angular momentum.

Thus we get

$$\psi_\alpha(x) \to \psi'_\alpha(x) = \exp\left[-\frac{i\theta}{\hbar}\underline{n}\cdot(\underline{L}+\underline{S})\right]\psi_\alpha(x) \tag{3.139}$$

$$= \left[\exp\left(-\frac{i\theta}{2}\underline{n}\cdot\underline{\sigma}\right)\right]^{\beta}_{\alpha}, \quad \psi_\beta(R^{-1}\underline{x}) \tag{3.140}$$

and we will just concentrate on the matrix factor.

Notice that

$$\underline{n}\cdot\underline{\sigma} \quad \underline{n}\cdot\underline{\sigma} = n_i n_j \sigma_i \sigma_j \tag{3.141}$$

$$= n_i n_j\{\delta_{ij}1 + i\varepsilon_{ijk}\sigma_k\} \tag{3.142}$$

$$= n^2 = 1 \tag{3.143}$$

so that

$$\exp\left(-\frac{i\theta}{2}\underline{n}\cdot\underline{\sigma}\right) = \cos\left(\frac{\theta}{2}\underline{n}\cdot\underline{\sigma}\right) - i\sin\left(\frac{\theta}{2}\underline{n}\cdot\underline{\sigma}\right) \tag{3.144}$$

$$= \cos\left(\frac{\theta}{2}\right) - i\underline{n}\cdot\underline{\sigma}\sin\left(\frac{\theta}{2}\right) \tag{3.145}$$

$$\therefore \psi_\alpha(\underline{x}) \to \left[\cos\frac{\theta}{2} - i\underline{n}\cdot\underline{\sigma}\sin\left(\frac{\theta}{2}\right)\right]^{\beta}_{\alpha}\psi_\beta(R^{-1}(\theta)\underline{x}). \tag{3.146}$$

We first introduced our angle θ to rotate a vector, and we expected a rotation of 2π to take us back to the start. Here we see that

$$\left(\psi(\underline{x} \overset{\theta=2\pi}{\to} (-)\psi(\underline{x})\right) \tag{3.147}$$

and we seem to have to go round twice in ordinary space to get back to $\psi(\underline{x})$ again. No physics goes wrong, because we always have bilinears in ψ to describe physical quantities. This is what distinguishes spinor "representations" of $SO(3)$ from vectors. (The spinors of $SO(3)$ are vectors of $SU(2)$.)

To check that we are not missing the point, we should do the finite angle transformations of both spinors and vectors in a full and related way. We shall need some more powerful machinery.

Orbital Angular Momentum ($\underline{x} \times \underline{p}$)

This is where the story usually starts. One way to represent the angular momentum algebra

$$[L_i, L_j] = i\varepsilon_{ijk}L_k \tag{3.148}$$

is to consider position and momentum vectors in a [3] space, with an uncertainty relation

$$[x_i, p_j] = i\delta_{ij}(\hbar) \tag{3.149}$$

$$L_i = \varepsilon_{ijk}x_j p_k. \tag{3.150}$$

Then

$$[L_i, L_j] = \varepsilon_{ipq}\varepsilon_{jlm}[x_p p_q, x_l p_m] \tag{3.151}$$

$$= \varepsilon_{ipq}\varepsilon_{jlm}\{x_p[p_q, x_l]p_m + x_l[x_p, p_m]p_q\} \tag{3.152}$$

$$= i\varepsilon_{ipq}\varepsilon_{jlm}\{-x_p p_m\delta_{ql} + x_l p_q\delta_{pm}\} \tag{3.153}$$

$$= ix_a p_b\{-\varepsilon_{ias}\varepsilon_{jsb} + \varepsilon_{isb}\varepsilon_{jas}\} \tag{3.154}$$

$$= ix_a p_b\left\{\delta_i^j\delta_a^b - \delta_i^b\delta_a^j - \delta_i^j\delta_a^b + \delta_i^a\delta_b^j\right\} \tag{3.155}$$

$$= ix_a p_b\varepsilon_{ijk}\varepsilon_{kab} \tag{3.156}$$

$$= i\varepsilon_{ijk}L_k \tag{3.157}$$

as required. Of course, you cannot represent all cases (add half-integers) this way, but we shall see this in detail later.

As you know, the Schrödinger representation is the one in which

$$\hat{x} \to x \tag{3.158}$$

$$\hat{p} \to -i(\hbar)\underline{\nabla} \tag{3.159}$$

but, although this specifies the concrete representation

$$L_i \to -i(\hbar)\varepsilon_{ijk}x_j\frac{\partial}{\partial x^k} \tag{3.160}$$

it does not reveal the full significance in three dimensions. To do this you move to *spherical polar* coordinates.

By thinking about small angles

$$\underline{e}_r = \sin\theta\cos\phi\underline{i} + \sin\theta\sin\phi\underline{j} + \cos\theta\underline{k} \tag{3.161}$$

$$\underline{e}_\theta = \cos\theta\cos\phi\underline{i} + \cos\theta\sin\phi\underline{j} - \sin\theta\underline{k} \tag{3.162}$$

$$\underline{e}_\phi = -\sin\phi\underline{i} + \cos\phi\underline{j}. \tag{3.163}$$

These $\underline{e}$'s are not fixed. An alternative (instructive) way to obtain these results is to note that

$$\underline{\nabla} = \underline{e}_r \frac{\partial}{\partial r} + \underline{e}_\theta \frac{1}{r} \frac{\partial}{\partial \theta} + \underline{e}_\phi \frac{1}{r \sin\theta} \frac{\partial}{\partial \phi}, \tag{3.164}$$

(watch the order) and to act on

$$\underline{r} = r\underline{e}_r = \{\sin\theta(\underline{i}\cos\phi + \underline{j}\sin\phi) + \underline{k}\cos\theta\}r \tag{3.165}$$

with $\underline{e}_\theta \cdot \underline{\nabla}$ and $\underline{e}_\phi \cdot \underline{\nabla}$.

Now, if you try to work out L_i in terms of spherical polars it soon gets hard. The example

$$\frac{\partial}{\partial \phi} = \frac{\partial x}{\partial \phi}\frac{\partial}{\partial x} + \frac{\partial y}{\partial \phi}\frac{\partial}{\partial y} + \frac{\partial z}{\partial \phi}\frac{\partial}{\partial z} \tag{3.166}$$

$$= -r\sin\theta\sin\phi\frac{\partial}{\partial x} + r\sin\theta\cos\phi\frac{\partial}{\partial y} + 0 \tag{3.167}$$

$$= -y\frac{\partial}{\partial x} + x\frac{\partial}{\partial y} \Rightarrow L_z = -i\frac{\partial}{\partial \phi} \tag{3.168}$$

is the only easy one. Try it and see. Instead you say:

$$\underline{L} = \underline{r} \times \underline{p} \Rightarrow -i\underline{r} \times \underline{\nabla} \tag{3.169}$$

$$= -ir\underline{e}_r \times \underline{\nabla} \qquad \{\underline{e}_r, \underline{e}_\theta, \underline{e}_\phi\} \quad \text{are right-hand set} \tag{3.170}$$

$$= -i\left[\frac{-\underline{e}_\theta}{\sin\theta}\frac{\partial}{\partial \phi} + \underline{e}_\phi\frac{\partial}{\partial \theta}\right] \tag{3.171}$$

$$= i\left\{\begin{array}{l} \underline{i}\left(\cot\theta\cos\phi\frac{\partial}{\partial\phi} + \sin\phi\frac{\partial}{\partial\theta}\right) \\ +\underline{j}\left(\cot\theta\sin\phi\frac{\partial}{\partial\phi} - \cos\phi\frac{\partial}{\partial\theta}\right) \\ +\underline{k}\left(-\frac{\partial}{\partial\phi}\right) \end{array}\right\}. \tag{3.172}$$

Hence

$$L_z = -i\frac{\partial}{\partial \phi}, \tag{3.173}$$

as before, while

$$L_\pm = L_x \pm iL_y = \pm e^{\pm i\phi}\left(\frac{\partial}{\partial \theta} \pm i\cot\theta\frac{\partial}{\partial \phi}\right). \tag{3.174}$$

Again,

$$\underline{L}^2 = L_i L_i = \frac{1}{2}(L_+ L_- + L_- L_+) + L_{z^2} \tag{3.175}$$

$$= L_z^2 + (\text{ Part of } L_+ L_- \text{ even in } \theta \to -\theta, \phi \to -\phi) \tag{3.176}$$

$$= -\frac{\partial^2}{\partial \phi^2} + \text{"Even part"}$$

$$\text{of } \left\{ e^{i\phi}\left(\frac{\partial}{\partial \theta} + i \cot\theta \frac{\partial}{\partial \phi}\right)(-)e^{-i\phi}\left(\frac{\partial}{\partial \theta} - i \cot\theta \frac{\partial}{\partial \phi}\right) \right\} \tag{3.177}$$

$$= -\frac{\partial^2}{\partial \phi^2} + \text{"Even part"}$$

$$\text{of } \left\{ \begin{array}{c} -\frac{\partial}{\partial \theta}\left(\frac{\partial}{\partial \theta} - i \cot\theta \frac{\partial}{\partial \phi}\right) \\ -e^{i\phi} i \cot\theta \left[\begin{array}{c} -ie^{-i\phi}\left(\frac{\partial}{\partial \theta} - i \cot\theta \frac{\partial}{\partial \phi}\right) \\ +e^{-i\phi}\frac{\partial}{\partial \phi}\left(\frac{\partial}{\partial \theta} - i \cot\theta \frac{\partial}{\partial \phi}\right) \end{array} \right] \end{array} \right\} \tag{3.178}$$

$$= -\frac{\partial^2}{\partial \phi^2} - \left\{ \begin{array}{c} \frac{\partial^2}{\partial \theta^2} \\ +\cot\theta \frac{\partial}{\partial \theta} \quad +\cot^2\theta \frac{\partial^2}{\partial \phi^2} \end{array} \right\} \tag{3.179}$$

$$= -\left\{ \frac{\partial^2}{\partial \theta^2} + \cot\theta \frac{\partial}{\partial \theta} + \frac{1}{\sin^2\theta}\frac{\partial^2}{\partial \phi^2} \right\} \tag{3.180}$$

$$= (-)\left\{ \frac{1}{\sin\theta}\frac{\partial}{\partial \theta}\sin\theta \frac{\partial}{\partial \theta} + \frac{1}{\sin^2\theta}\frac{\partial^2}{\partial \phi^2} \right\}. \tag{3.181}$$

Note: The $\{\underline{e}_r, \underline{e}_\theta, \underline{e}_\phi\}$ are not fixed, so to get this result by $(-i\underline{r} \times \underline{\nabla}) \cdot (-i\underline{r} \times \underline{\nabla})$ in spherical polars needs $\underline{\nabla} \cdot \underline{e}_\theta$ and so forth. Try if you like; it is not easy.

Now, what we have shown is that L_i does not depend upon r at all. The action of the L_i on the space is to induce rotations. We shall return to this later. Meantime, we ought to specify the representations. You should solve

$$L_z Y_\rho^m(\theta, \phi) = m Y_l^m(\theta, \phi) \tag{3.182}$$

$$L^2 Y_\rho^m(\theta, \phi) = l(l+1) Y_l^m(\theta, \phi), \tag{3.183}$$

but L^2 is horrible and has two derivatives, so you solve instead

$$L_z Y_l^m(\theta, \phi) = m Y_l^m(\theta, \phi) \tag{3.184}$$

$$L_\pm Y_l^m(\theta, \phi) = \sqrt{(l \mp m)(l \pm m + 1)} Y_l^m(\theta, \phi). \tag{3.185}$$

$$L_z Y_l^m = m Y_l^m \tag{3.186}$$

$$\Rightarrow \frac{\partial}{\partial \phi} Y_l^m = i m Y_l^m \tag{3.187}$$

$$\therefore \quad Y_l^m = c_{lm} e^{im\phi} P_l^m(\theta). \tag{3.188}$$

$$L_+ Y_l^l = 0 \tag{3.189}$$

$$\Rightarrow e^{i\phi} \left(\frac{\partial}{\partial \theta} + i \cot\theta \frac{\partial}{\partial \phi} \right) e^{il\phi} P_l^l(\theta) = 0 \tag{3.190}$$

$$\therefore \quad \left(\frac{\partial}{\partial \theta} - l \cot\theta \right) P_l^l(\theta) = 0 \tag{3.191}$$

$$\therefore \quad \frac{d P_l^l}{P_l^l} = l \cot\theta d\theta = l \frac{d(\sin\theta)}{\sin\theta} \tag{3.192}$$

$$\therefore \quad P_l^l = (\text{const.}) \sin^l \theta. \tag{3.193}$$

We know (by the orthogonality theorem) that these functions can be made orthonormal, and we normalize on the unit sphere

$$\int_0^\pi d\theta \sin\theta \int_0^{2\pi} d\phi Y_{\varrho'}^{*m'}(\theta, \phi) Y_l^m(\theta, \phi) = \delta_{ll'} \delta_{mm'}. \tag{3.194}$$

Here we need

$$1 = |c_{ll}|^2 \int_0^{2\pi} d\phi \int_0^\pi d\theta \sin\theta \sin^{2l}\theta \tag{3.195}$$

$$= |c_{ll}|^2 2\pi \int_0^\pi d\theta \sin^{2l+1}\theta$$

$$= |c_{ll}|^2 2\pi \, I_{2l+1}. \tag{3.196}$$

But

$$I_{2l+1} = -\int \sin^{2l}\theta d(\cos\theta) \tag{3.197}$$

$$= -[\cos\theta \sin^{2l\theta}]_0^\pi + \int_0^\pi \cos\theta d(\sin^{2l}\theta) \tag{3.198}$$

$$= \int_0^\pi 2l \cos^2\theta \sin^{2l-1} d\theta \tag{3.199}$$

$$= 2l\, I_{2l-1} - 2l\, I_{2l+1} \tag{3.200}$$

$$\therefore \quad I_{2l+1} = \frac{2l}{2l+1} I_{2l-1} = \frac{(2l)!!}{(2l+1)!!} I_1$$

$$= \frac{(2l!!)^2}{(2l+1)!} I_1 \tag{3.201}$$

$$= \frac{(2^l l!)^2}{(2l+1)!} 2 = \frac{2^{l+1}(l!)^2}{(2l+1)!}. \tag{3.202}$$

Convention then takes c_{ll} real, and the sign $(-1)^l$ to make $Y_l^0(0,0)$ real and positive, and we find

$$Y_l^l = \frac{(-1)^l}{2^l l!} \sqrt{\frac{(2l+1)!}{4\pi}} e^{il\phi} \sin^l\theta. \tag{3.203}$$

The rest are now found by lowering:

$$Y_l^{m-1} = \sqrt{\frac{1}{(l+m)(l-m+1)}} L_- Y_l^m \tag{3.204}$$

$$\therefore \quad Y_l^m = \sqrt{\frac{(l+m)!}{2l!(l-m)!}} (L_-)^{l-m} Y_l^l. \tag{3.205}$$

The conventional connection between the spherical harmonics (Y_l^m) and the associated (first kind) Legendre functions (P_l^m) is:

$$Y_l^m = (-1)^m \sqrt{\frac{(2l+1)(l-m)!}{4\pi(l+m)!}} e^{im\phi} P_l^m \qquad (m \geq 0), \tag{3.206}$$

and in particular,

$$Y_l^0 = \sqrt{\frac{2l+1}{4\pi}} P_l \tag{3.207}$$

where the P_l are the Legendre polynomials (the 0 index is conventionally dropped).

Now

$$Y_0^0 = \sqrt{\frac{1}{4\pi}}, \tag{3.208}$$

$$Y_1^1 = -\sqrt{\frac{3}{8\pi}} e^{i\phi} \sin\theta \tag{3.209}$$

$$Y_1^0 = \sqrt{\frac{3}{4\pi}} \tag{3.210}$$

$$Y_1^{-1} = \sqrt{\frac{3}{8\pi}} e^{-i\phi} \sin\theta. \tag{3.211}$$

The Spinors Revisited

Recall that

$$U_{\alpha\beta} = \left[\exp\left(-\frac{i}{2}\theta\right) n_i \sigma_i\right]_{\alpha\beta} \tag{3.212}$$

$$\psi_\alpha \to U_{\alpha\beta}\psi_\beta \tag{3.213}$$

$$V_i \to R_{ij} V_j \tag{3.214}$$

are the transformation laws under an angle of magnitude θ about the axis parallel to $\underline{n}$.

The projection operators

$$P^{\pm} = \frac{1}{2}(1 \pm \underline{n} \cdot \underline{\sigma}) \tag{3.215}$$

have the usual properties in the spinor sector, so that

$$U = 1 \cos\left(\frac{\theta}{2}\right) - i\underline{\sigma} \cdot \underline{n} \sin\left(\frac{\theta}{2}\right) \tag{3.216}$$

as previously.

What are the related projection operators for the vector representation? Consider

$$(P^{AB})_{ij} = \frac{1}{2} Tr \cdot (P^A \sigma_i P^B \sigma_j) \qquad A \neq B \tag{3.217}$$

$$(I)_{ij} = \frac{1}{2} \sum_A Tr \cdot (P^A \sigma_i P^A \sigma_j) \tag{3.218}$$

where A takes the range 1 to 2, and i takes the range 1 to 3.

Use the completeness

$$(\sigma_i)_{\alpha\beta}(\sigma_i)_{\gamma\delta} + \delta_{\alpha\beta}\delta_{\gamma\delta} = 2\delta_{\alpha\delta}\delta_{\gamma\beta} \tag{3.219}$$

to write

$$\frac{1}{2}Tr\cdot(\sigma_i X)Tr\cdot(\sigma_i Y) = Tr\cdot(XY) - \frac{1}{2}Tr\cdot(X)Tr\cdot(Y) \tag{3.220}$$

$$\frac{1}{2}Tr\cdot(\sigma_i X\sigma_i Y) = Tr\cdot(X)Tr\cdot(Y) - \frac{1}{2}Tr\cdot(XY). \tag{3.221}$$

Then

$$P^{AB})_{ik}(P^{CD})_{kj} = \frac{1}{2}Tr\cdot\left(P^A\sigma_i P^B\sigma_k\right)\frac{1}{2}Tr\cdot\left(P^C\sigma_k P^D\sigma_j\right) \tag{3.222}$$

$$= \sum_{\text{no sum}} \frac{1}{2}Tr\cdot\left(P^A\sigma_i P^B\sigma_j\right)\delta^{AC}\delta^{BD} \tag{3.223}$$

by Equation 3.220

$$= 0 \text{ if } A \neq B \text{ and/or } C \neq D. \tag{3.224}$$

Again

$$(P^{AB})_{ik}I_{kj} = \frac{1}{2}Tr\cdot\left(P^A\sigma_i P^B\sigma_k\right)\frac{1}{2}Tr\cdot\left(P^C\sigma_k P^D\sigma_j\right) \tag{3.225}$$

$$= \text{by Equation (3.220)}$$

$$= 0 \text{ because } B \neq C \text{ etc.,} \tag{3.226}$$

and similarly $IP = 0$.
Yet again

$$I_{ik}I_{kj} = \frac{1}{2}Tr\cdot\left(P^A\sigma_i P^A\sigma_k\right)\frac{1}{2}Tr\cdot\left(P^B\sigma_k P^B\sigma_j\right) \tag{3.227}$$

$$= \text{ by Equation 3.220, } \frac{1}{2}Tr\cdot\left(P^A\sigma_i P^A\sigma_j\right) = I_{ij} \tag{3.228}$$

because $\sum P^A = 1$ and the σ_i are traceless.

Finally

$$\sum_{A\neq B}(P^{AB})_{ij} + I_{ij} = \delta_{ij} \tag{3.229}$$

so these are projection operators.

Dimensions of Projected Spaces

These are given by traces. $Tr \cdot (P^{AB})$ = by Equation (3.221) = 1, so this is one dimensional and there are 2(2 – 1) = 2 of these projectors.

$$Tr \cdot (I) = \frac{1}{2} Tr \cdot \left(P^A \sigma_i P^A \sigma_i\right) \tag{3.230}$$

$$= Tr \cdot (P^A) Tr \cdot (P^A) - \frac{1}{2} Tr \cdot (P^A P^A) \tag{3.231}$$

by Equation 3.221,

$$= \sum_{A=1}^{2} (1 \times 1) - \frac{1}{2} \sum_{A=1}^{2} (1) = 2 - 1 = 1 \tag{3.232}$$

as expected, but this space is not irreducible.

In our $SU(2)$ notation we find

$$(P^{\pm\mp})_{ij} = \frac{1}{2} Tr \cdot (P^{\pm} \sigma_i P^{\mp} \sigma_j) \tag{3.233}$$

$$= \frac{1}{2} [\delta_{ij} - n_i n_j \mp i \varepsilon_{ikj} n_k] \tag{3.234}$$

$$\text{and} \quad (I)_{ij} = \frac{1}{2} Tr \cdot [(P^+ \sigma_i P^+ \sigma_j) + (P^- \sigma_i P^- \sigma_j)] \tag{3.235}$$

$$= \frac{1}{2} Tr \cdot (\sigma_i \sigma_j) - (P^{+-} + P^{-+})_{ij} \tag{3.236}$$

$$= \delta_{ij} - (\delta_{ij} - n_i n_j) = n_i n_j. \tag{3.237}$$

Connection between the "Mixed Spinor" and the Adjoint (Regular) Representation

Recall that the product of two spin $\frac{1}{2}$ states gave a three-dimensional vector and a singlet. When the singlet has been removed by the trace, we can write

$$V_{\alpha\beta} = \frac{1}{\sqrt{2}} V_i (\sigma_i)_{\alpha\beta} \tag{3.238}$$

and the inverse

$$V_i = \frac{1}{\sqrt{2}} Tr \cdot (V \sigma_i). \tag{3.239}$$

Now when the spinor transforms as

$$\psi_\alpha \to U_{\alpha\beta}\psi_\beta \tag{3.240}$$

then we see that

$$V_i \to \frac{1}{\sqrt{2}} Tr \cdot (UVU^{-1}\sigma_i) \tag{3.241}$$

is the transformation of the vector. Expanding for small angles we see that

$$U = 1 - \frac{1}{2} i\underline{\theta} \cdot \underline{\sigma} \tag{3.242}$$

$$U^{-1} = 1 + \frac{1}{2} i\underline{\theta} \cdot \underline{\sigma}, \tag{3.243}$$

$$\to \frac{1}{\sqrt{2}} Tr \cdot (VU^{-1}\sigma_i U) \tag{3.244}$$

$$= V_i + \theta_k \varepsilon_{ikj} V_k \tag{3.245}$$

to first order, which we hope is familiar. We would like the full version of all this. Returning to

$$V_i \to \frac{1}{\sqrt{2}} Tr \cdot (UVU^{-1}\sigma_i) \tag{3.246}$$

we use Equation (3.220) from the completeness relations to establish that

$$V_i \to R_{ij} V_j \tag{3.247}$$

$$R_{ij} = \frac{1}{2} Tr \cdot (U^{-1}\sigma_i U \sigma_j). \tag{3.248}$$

It is straightforward to use Equation (3.220) again to show that

$$R_{ij}(R^T)_{jk} = \delta_{jk}, \tag{3.249}$$

where the T indicates transposition of the matrix, so that R is, as expected, actually orthogonal.

Finite Angle Rotation of $SO(3)$ Vector

We have just established that

$$U = \exp\left[-i\frac{\theta}{2}P^+ + i\frac{\theta}{2}P^-\right] \tag{3.250}$$

and that

$$R_{ij} = I_{ij} + P_{ij}^{+-} \exp\left[\frac{i}{2}\{(\theta) - (-\theta)\}\right] + (P_{ij}^{-+})\exp\left[\frac{i}{2}\{(-\theta) - (-\theta)\}\right] \tag{3.251}$$

$$= n_i n_j + (\delta_{ij} - n_i n_j)\cos\theta - i\varepsilon_{ikj} n_k i \sin\theta \tag{3.252}$$

$$\therefore \quad R(\underline{n}, \theta)_{ij} = n_i n_j[1 - \cos(\theta)] + \delta_{ij}\cos(\theta) + \varepsilon_{ikj} n_k \sin\theta. \tag{3.253}$$

This is often written in the form of dot and cross products as

$$\underline{V} \to \underline{V}' = \underline{V}\cos(\theta) + [1 - \cos(\theta)](\underline{V} \cdot \underline{n}) + \sin(\theta)\underline{n} \times \underline{V}. \tag{3.254}$$

(If you are not familiar with this result, it could be checked by using the [cubic] characteristic equation method.) Note that when $\theta \to 2\pi$ then $\underline{V}' \to V$. So the rotations really do what we expect and the spinors of $SO(3)$ really do a strange double cover transformation.

References

1. A.R. Edwards, *Angular Momentum in Quantum Mechanics*. Princeton University Press, Princeton, 1957.
2. M.E. Rose, *Elementary Theory of Angular Momentum*. John Wiley & Sons, New York, 1957.

Problems

3.1 Read through from Equations (3.4) to (3.8). Now close the book and try to rewrite them.

3.2 Explain what is understood by a contravariant vector.

3.3 Explain what is understood by a covariant vector.

3.4 Show how to express a tensor T^{ij} into symmetric and antisymmetric parts. Using the Levi-Civita tensor show how to write the antisymmetric part as a covariant vector.

3.5 Taking a contravariant vector as the example, show how transitivity works.

3.6 Using your own notation, show that for rotations in three dimensions, covariant and contravariant are equivalent.

3.7 Explain in your own words what is understood by a scalar operator and a vector operator.

3.8 Show in terms of Dirac's state vectors that operators representing transformations on the space formed by them are unitary.

3.9 Show how to express the operator $R(\underline{e}_3, \theta)$ as a 3×3 matrix.

3.10 Using your own notation show how to build up a rotation operator in three dimensions by an angle about a fixed vector axis.

3.11 Take the rotation operator you constructed in the previous problem and by rotating a vector operator K_i and then expanding to first nontrivial order in the angle, find the commutator of K_i with the components of the angular momentum. Hence, retrieve the familiar commutation relations of angular momentum.

3.12 Using your own notation show how to describe the angular momentum of a spin $\frac{1}{2}$ wave function. Use your result to find the 2×2 transformation on a spin $\frac{1}{2}$ wave function.

3.13 Using your own notation, show how to find the commutation relations of components of orbital angular momentum for the integer cases.

3.14 Using spherical polar coordinates, work out the operator L_z.

3.15 Repeat Problem 3.14 starting with $\underline{L} = \underline{r} \times \underline{p}$.

3.16 Using your own notation work out the correctly normalized form of the spherical harmonics Y_1^1, Y_1^0, and Y_1^{-1}.

3.17 Show that the familiar operators $P^{\pm} = \frac{1}{2}(1 \pm \underline{n} \cdot \underline{\sigma})$ are indeed projection operators.

3.18 Show what is understood by completeness of the Pauli matrices and the associated unit operator matrix.

3.19 What dimension do the state vectors projected by the operators in the previous two problems have?

3.20 The product of two spin $\frac{1}{2}$ states gives a three-dimensional vector and a singlet. When the singlet has been removed by tracing, show how the components of the three-dimensional vector can be expressed in terms of the trace of the matrix and Pauli matrices and also vice versa.

4

Special Relativity and the Physical Particle States

The Dirac Equation

We look for a Hamiltonian to use in the time-dependent Schrödinger equation, and because this is first order in time we stay first order in space trying to treat all coordinates on a similar basis. Write

$$i\frac{\partial\psi}{\partial t} = H\psi = -i\alpha^i\frac{\partial\psi}{\partial x^i} + m\beta\psi \quad i = 1, 2, 3 \tag{4.1}$$

where ψ is now a column and the α^i and β are square matrices.

We require $p^2 = m^2$ or the Klein–Gordon equation $(\Box + m^2)\psi = 0$ but we actually have

$$0 = \frac{\partial^2\psi}{\partial t^2} - \frac{\alpha^i\alpha^j + \alpha^j\alpha^i}{2}\frac{\partial^2\psi}{\partial x^i\partial x^j} - im(\alpha^i\beta + \beta\alpha^i)\frac{\partial\psi}{\partial x^i} + m^2\beta^2\psi \tag{4.2}$$

by iteration (or squaring) and using symmetry in the second term. Hence we require the algebra

$$\begin{aligned} \alpha^i\alpha^j + \alpha^j\alpha^i &= 2\delta^{ij} \\ \alpha^i\beta + \beta\alpha^i &= 0 \\ (\alpha^i)^2 = 1 &= \beta^2. \end{aligned} \tag{4.3}$$

Moreover if H is Hermitian, then so are α^i and β.

Now any matrix whose square is 1 has eigenvalues only ± 1. (Proof: $M\psi = a\psi$ with real a because M is Hermitian. Thus $\psi = M^2\psi = M a\psi = a^2\psi$.) Also β and α^i are traceless. (Proof: $Tr(\alpha^i) = Tr(\beta^2\alpha^i) = Tr(\beta\alpha^i\beta) = Tr(-\alpha^i)$.)

Hence since the trace of a matrix is the sum of the eigenvalues, there must be an equal number of ± 1's—these matrices are thus of even dimensions. Clearly 2×2 is too small—only three Pauli matrices but at least four Dirac matrices. We shall try 4×4.

All possible products will now therefore yield (up to) 16 independent matrices that are Hermitian. But for reasons that will become clear (covariance

DOI: 10.1201/9781439895207-4

with pseudo-orthogonal Lorentz group), we choose to work with 16 that have a pseudo-Hermitian property. Define

$$\gamma^0 = \beta \quad \text{and} \quad \gamma^i = \beta\alpha^i \tag{4.4}$$

so that the Dirac equation now reads

$$i\left\{\gamma^0 \frac{\partial}{\partial t} + \gamma' \frac{\partial}{\partial x'} + +\right\}\psi = m\psi$$

$$\text{or} \quad i\gamma^\mu \partial_\mu \psi = m\psi$$

$$\text{or} \quad i\not{\partial}\psi = m\psi \tag{4.5}$$

where we have written $\gamma^\mu p_\mu = \not{p} = p^\mu \gamma_\mu$ for p^μ any vector.

Notes: (a) If $p^\mu = i\partial^\mu \rightarrow (\not{p} - m)\psi = 0$. (b) Despite the index notation the γ^μ are just four matrices (similarly $\gamma_\mu = g_{\mu\nu}\gamma^\nu$) and they never transform.

The Clifford Algebra: Properties of γ Matrices

The *Clifford algebra* now reads

$$\{\gamma^\mu, \gamma^\nu\} = 2g^{\mu\nu} \tag{4.6}$$

where 1 has been suppressed on the right-hand side. All the γ's are traceless. γ^0 is Hermitian and the γ^i are anti-Hermitian. We have four of the γ's and the unit matrix so there are 11 more objects to find. First look at the commutators of the γ's:

$$\sigma^{\mu\nu} = \frac{i}{2}[\gamma^\mu, \gamma^\nu] \tag{4.7}$$

and this gives us six more because of the antisymmetry in $\mu \leftrightarrow \nu$.

Clearly the σ^{ij} are Hermitian and the σ^{0i} anti-Hermitian. All $\sigma^{\mu\nu}$ are traceless because any product of anticommuting matrices is traceless. (Proof: $\mathrm{Tr}(AB) \underset{\text{anti}}{=} (-)Tr(BA) \underset{\text{cyclic}}{=} (-)Tr(AB)$. This assumes finiteness.)

Now because each of the γ^μ squares to 1, we can see that the last five matrices must be four of the type $\gamma^0\gamma^1\gamma^2$ (i.e., missing one of γ^μ each time) and with $\gamma^0\gamma^1\gamma^2\gamma^3$ as the final one. We define

$$\sigma^{\mu 5} = i\gamma^\mu\gamma^5 = \frac{i}{3!}\varepsilon^{\mu\nu\varrho\lambda}\gamma_{-}\gamma_\varrho\gamma_\lambda \tag{4.8}$$

where using $\gamma_\mu\gamma^\mu = 4$

$$\gamma^5 = \frac{\varepsilon^{\mu\nu\varrho\lambda}}{4!}\gamma_\mu\gamma_\nu\gamma_\varrho\gamma_\lambda. \tag{4.9}$$

Notice that $\varepsilon^{0123} = (-)1$ so that $\gamma^5 = (-)\gamma_0\gamma_1\gamma_2\gamma_3 = \gamma^0\gamma^1\gamma^2\gamma^3 = \gamma^3\gamma^2\gamma^1\gamma^0$ and so forth. Clearly, by looking at the 3γ and 4γ expressions, σ^{i5} are Hermitian, but γ^5 and σ^{05} are anti-Hermitian. Again, $\{\gamma^\mu, \gamma^5\} = \{\gamma^\mu, \gamma^0\gamma^1\gamma^2\gamma^3\}$

and with three of these γ^μ anticommutes but with itself it commutes so that

$$\{\gamma^\mu, \gamma^5\} = 0. \tag{4.10}$$

Thus

$$\sigma^{\mu 5} = \frac{i}{2}[\gamma^\mu, \gamma^5], \tag{4.11}$$

which fits with the previous notation and also shows that $\mathrm{Tr}(\sigma^{\mu 5}) = 0$ because it is the product of anticommuting matrices.

Finally $\gamma^5 = \gamma^0(\gamma^1\gamma^2\gamma^3) = -(\gamma^1\gamma^2\gamma^3)\gamma^0$ so that γ^5 is traceless too. Moreover

$$\gamma^5\gamma^5 = \gamma^0\gamma^1\gamma^2\gamma^3\gamma^3\gamma^2\gamma^1\gamma^0 = -1$$
$$\therefore \{\gamma^5, \gamma^5\} = -2. \tag{4.12}$$

You now have 16 matrices, all Hermitian or anti-Hermitian, and all traceless except for one.

To establish that they are linearly independent you work out the full multiplication table (Figure 4.1).

	B
A	½ $\vert A, B\vert$
	½ $\{A, B\}$

	1	γ^ρ	$\sigma^{\rho\lambda}$	$\sigma^{\rho 5}$	γ^5
1	0	0	0	0	0
	1	γ^ρ	$\sigma^{\rho\lambda}$	$\sigma^{\rho 5}$	γ^5
γ^μ		i	$i\vert g^{\mu\rho}\gamma^\lambda - g^{\mu\lambda}\gamma^\rho\vert$	$i\, g^{\mu\rho}\gamma^5$	$-i\,\sigma^{\mu 5}$
		$-g^{\mu\rho}$	$\varepsilon^{\mu\rho\lambda\delta}\sigma_{\delta 5}$	$-½\,\varepsilon^{\mu\rho\alpha\beta}\sigma_{\alpha\beta}$	0
$\sigma^{\mu\nu}$			$i\vert\sigma^{\mu\lambda}g^{\nu\rho} - \sigma^{\mu\rho}g^{\nu\lambda} + \sigma^{\lambda\nu}g^{\mu\rho} - \sigma^{\rho\nu}g^{\mu\lambda}\vert$	$-i\vert g^{\mu\rho}\sigma^{\nu 5} - g^{\rho\nu}\sigma^{\rho 5}\vert$	0
			$g^{\mu\rho}g^{\mu\lambda} - g^{\mu\lambda}g^{\nu\rho} + \varepsilon^{\mu\nu\rho\lambda}\gamma^5$	$\varepsilon^{\mu\nu\rho\delta}\gamma_\delta$	$-½\,\varepsilon^{\mu\nu\alpha\beta}\sigma_{\alpha\beta}$
$\sigma^{\mu 5}$				$i\sigma^{\mu\rho}$	$-i\,\gamma^\mu$
				$-g^{\mu\rho}$	0
γ^5					0
					-1

FIGURE 4.1
The γ matrix product table.

Structure of the Clifford Algebra and Representation

If you examine the table (Figure 4.1) you see that the square of each of the $16\Gamma^{\mu}$ is ± 1. Indeed, you can see that $(\Gamma^R)^{\dagger} \equiv (\Gamma^R)^{-1}$ for all R. Moreover the product of any two classes

$$\Gamma^R\Gamma^S = p^{RSQ}\Gamma^Q \quad (p \text{ are complex numbers; each is } \pm 1 \text{ or } \pm i) \tag{4.13}$$

in such a way that there is only one term in the sum, and only one appears when you have a square.

1. The 16 Γ^R give a linearly independent basis.
   Proof: If $c_R\Gamma^R = 0$ then multiply by Γ^S and trace $\to 0 = c_R\mathrm{Tr}(\Gamma^S\Gamma^R) = \Sigma_S 4\xi_S c_S \to c_S \equiv 0$ where we have defined $(\Gamma^R)^2 = \xi_R 1$. We see that

$$\Sigma_S \xi_S \Gamma^S = (\Gamma^S)^{-1} = (\Gamma^S)^{\dagger}, \qquad \xi_S = \pm 1 \tag{4.14}$$

   is another way of writing this.

2. $$\Sigma_R \xi_R p^{SRQ} = \Sigma_Q \xi_Q p^{QSR}$$

$$\text{Proof:} \quad \Gamma^Q\Gamma^S = p^{QSR}\Gamma^R$$

$$(\Gamma^Q \times) \quad \xi_Q\Gamma^S = \Sigma_\alpha p^{QSR}\Gamma^Q\Gamma^R$$

$$(\times\Gamma^R) \quad \xi_Q\Gamma^S\Gamma^R = \Sigma_{Q,R} p^{QSR}\Gamma^Q\xi_R$$

   (There was only one term in R "sum.")

$$(\xi_R\xi_Q) \quad \Sigma_R \xi_R \Gamma^S\Gamma^R = \Sigma_Q \xi_Q p^{QSR}\Gamma^Q$$

$$\therefore \Sigma_R \Sigma_Q \xi_R p^{SRQ}\Gamma^Q = \Sigma_\alpha \xi_Q p^{QSR}\Gamma^Q$$

$$\therefore \Sigma_R \xi_R p^{SRQ} = \Sigma_Q \xi_Q p^{QSR}. \tag{4.15}$$

3. If there are two representations of the algebra of multiplication, Γ^R, which is $N \times N$, and γ^R, which is $n \times n$ then there is a matrix linking them.
   Proof: Let F be an arbitrary $N \times n$ matrix and define $S = \xi_R\Gamma^R F\gamma^R$ where the triple index is intended and there are 16 terms in the sum. Then

$$\begin{aligned} S\gamma^S &= \xi_R\Gamma^R F p^{RSQ}\gamma^Q \\ &= \xi_Q\Gamma^Q F p^{QSR}\gamma^R \quad \text{(relabel)} \\ &= \xi_R p^{SRQ}\Gamma^Q F\gamma^R \quad \text{(by 2 above)} \\ &= \Gamma^S\xi_R\Gamma^R F\gamma^R \quad \text{(same algebra for } \gamma \text{ and } \Gamma) \\ &= \Gamma^S S. \end{aligned} \tag{4.16}$$

Notice that F is quite arbitrary and since the γ's are linearly independent we can always find an S which is not identically zero (e.g., take just one entry of $F \neq 0$).

4. Up to equivalence there is only one irreducible representation of the γ matrices. Up to a constant factor the matrix giving the equivalence is unique.

Proof: We have to invoke Schur's lemma; we will prove this on the way. The first part of Schur's lemma states that if the representations are irreducible (i.e., there are no nontrivial subspaces; the identity and the whole lot are trivial) then S is either null, or is nonsingular with $N = n$.

To prove this note that Γ and γ induce linear transformations in the vector spaces P_N and P_n. But S maps P_n into a subspace (we assume $N \geq n$ without loss of generality) $\tilde{P}$ of P_N by

$$S : P_n \to \tilde{P} = sP \subset P_N \qquad (S \text{ is } N \times n)$$

and this $\tilde{P}$ is an invariant subspace of P_N because

$$\Gamma^S\{\tilde{P}\} = \Gamma^S S\{P\} = S\gamma^s\{P\} = \text{ set multiplication } S\{P\} = \tilde{P}$$

so the irreducibility implies $\tilde{P}$ is either empty or identical to P_N. But we know that S is not null, so $P_N = S\{P_N\}$ is a subspace of P_n while $n \leq N$; clearly $n = N$. Moreover S has an inverse, because if it had zero determinant it would pick out a subspace. More precisely, if S is singular then at least one eigenvalue is zero, so let $\{P^0\}$ be the set with $S\{P^0\} = 0$. Then

$$S\gamma^R\{P^0\} = \Gamma^R S\{P^0\} = \Gamma^R 0 = 0$$

$$\gamma^R\{P^0\} = \{P^0\}$$

and $\{P^0\}$ is an invariant subspace.

But S is not null (which could be the case if all the eigenvalues were zero), hence we have a contradiction, so S is not singular.

The two representations are equivalent by

$$\Gamma^S = S\gamma^S S^{-1} \tag{4.17}$$

and we can now show that S is unique up to a multiplicative factor by using the second part of Schur's lemma: If a matrix commutes with all members of the irreducible set then it must be a multiple of the identity. The proof of this starts from $[\gamma^R, M] = 0$ so that $[\gamma^R, M - \lambda 1] = 0$.

But take λ such that $\text{Det}(M - \lambda 1) = 0$, then it picks out a subspace unless it is null, so Schur's lemma is saying $M = \lambda 1$.

So now, if you have two matrices S and S', which both give the equivalence

$$\Gamma^S A = S\gamma^S S^{-1}$$
$$\Gamma^S = S'\gamma^S S'^{-1}$$
$$\text{then} \quad S^{-1}S'\gamma^S S'^{-1}S = S^{-1}\Gamma^S S = \gamma^S$$
$$\therefore [S^{-1}S', \gamma^S] = 0$$
$$\therefore S^{-1}S' = \lambda 1$$
$$\therefore S' = \lambda S.$$

5. Completeness. Consider the section above when $\Gamma^R = \gamma^R$ and therefore $S = \lambda 1$.

$$\text{Then} \quad \lambda\delta_\alpha^\beta = S_\alpha^\beta = \xi_R(\gamma^R)_\alpha^\nu (F_\nu^\mu)(\gamma^R)_\mu^\beta$$
$$\text{and tracing} \quad \Rightarrow 4\lambda = \xi_R(\gamma^R)_\alpha^\nu F_\nu^\mu(\gamma^R)_\mu^\alpha$$
$$= \xi_R\xi_R\delta_\mu^\nu F_\nu^\mu$$
$$= 16\delta_\mu^\nu F_\nu^\mu.$$

Hence $\lambda = 4\delta_\mu^\nu F_\nu^\mu$ and substituting

$$4\delta_\mu^\nu F_\nu^\mu \delta_\alpha^\beta = \xi_R(\gamma^R)_\alpha^\nu F_\nu^\mu(\gamma^R)_\mu^\beta$$

and this is true for arbitrary F implies

$$\sum_{R=1}^{1} 6\xi_R(\gamma^R)_\alpha^\nu(\gamma^R)_\mu^\beta = 4\delta_\alpha^\beta\delta_\mu^\nu. \tag{4.18}$$

Lorentz Covariance of the Dirac Equation

Recall that the rotations and boosts are specified by the coordinates of the transformed point being given as

$$x'^\mu = \Omega_\nu^\mu x^\nu$$
$$\partial'_\mu = \frac{\partial}{\partial x'^\mu} = \Omega_\mu^\nu \partial_\nu.$$

Now the Dirac equation reads $(i\gamma^\mu\partial_\mu - m)\psi(x) = 0$ and if $\psi \to \psi'$ under Lorentz transformation, then (examining at the transformed point for convenience) we had better have $(i\gamma^\mu\partial'_\mu - m)\psi'(x') = 0$ to ensure covariance. We can rewrite this as $(i\gamma^\mu\Omega_\mu^\nu\partial_\nu - m)\psi'(\Omega x) = 0$ and writing $\Gamma^\nu = \gamma^\mu\Omega_\mu^\nu$ we can insist that $\Gamma^\nu = s_\Omega \gamma^\nu s_\Omega^{-1}$ because

$$\begin{aligned}\{\Gamma^\mu, \Gamma^\nu\} &= \left\{\gamma^\alpha \Omega^\mu_\alpha, \gamma^\beta \Omega^\nu_\beta\right\} \\ &= \Omega^\mu_\alpha \Omega^\nu_\beta 2g^{\alpha\beta} \\ &= 2g^{\mu\nu}\end{aligned}$$

and we have the same algebra and can apply the theorem. (Of course we do not yet know S.) We now have $(i\gamma^\mu\partial_\mu - m)S_\Omega^{-1}\psi'(\Omega x) = 0$ since S is nonsingular. So, to ensure covariance, we have

$$\psi(x) \to \psi'(x) = S_\Omega \psi(\Omega^{-1}x) \tag{4.19}$$

under Lorentz transformation. Any object that transforms in this manner is called a covariant *four-component spinor*. Of course, the S_n have to form a representation of the Lorentz group.

Now, using Lie's theorems again, we restrict to the infinitesimal form $\Omega^\nu_\mu = \delta^\nu_\mu + \omega^\nu_\mu + \ldots$ and for the spinors we write

$$\begin{aligned}\psi'_\alpha(\Omega x) &= S^\beta_\alpha \psi_\beta(x) \\ &= \psi_\alpha(x) - \frac{i}{4}\omega_{\mu\nu}(\Sigma^{\mu\nu})^\beta_\alpha \psi_\beta(x) + \ldots \end{aligned}\tag{4.20}$$

where the six $\Sigma^{\mu\nu}$ have to be constructed out of the 16 matrices Γ^R.

Imposing

$$\gamma^\rho \Omega^\lambda_\rho = S\gamma^\lambda S^{-1}$$

implying that

$$\begin{aligned}\gamma^\rho[\delta^\lambda_\rho + \omega^\lambda_\rho] &\simeq \left[1 - \frac{i}{4}\underline{\omega}\cdot\underline{\Sigma}\right]\gamma^\lambda\left[1 + \frac{i}{4}\underline{\omega}\cdot\underline{\Sigma}\right] \\ \therefore \gamma^\rho\omega^\lambda_\rho &\simeq -\frac{i}{4}\omega_{\mu\nu}[\Sigma^{\mu\nu}, \gamma^\lambda]\end{aligned}$$

that is,

$$\omega_{\mu\nu}\left\{[\Sigma^{\mu\nu}, \gamma^\lambda] - 2i[\gamma^\mu g^{\nu\lambda} - \gamma^\nu g^{\mu\lambda}]\right\} = 0.$$

But the $\omega_{\mu\nu}$ are arbitrary, so that

$$[\Sigma^{\mu\nu}, \gamma^\lambda] = 2i(\gamma^\mu g^{\nu\lambda} - \gamma^\nu g^{\mu\lambda}) \tag{4.21}$$

and from the tables $\Sigma^{\mu\nu} = \sigma^{\mu\nu}$ is a solution of this. Notice that for proper Lorentz transformations we require $\det S = 1$ so that the $\sigma^{\mu\nu}$, being traceless, are appropriate (i.e., do not take the solution $\sigma^{\mu\nu} + 1$). Also, if there is another solution, then the difference commutes with all $\gamma^\lambda \Rightarrow$ commutes all $\Gamma^R \Rightarrow$ is a multiple of the identity. So we have the solution.

The Adjoint

We know that some of our Γ^R are Hermitian and others anti-Hermitian. Notice, by inspection, that

$$\overline{\gamma^R} \equiv \gamma^0 \gamma^{R\dagger} \gamma^0 = \gamma^R \tag{4.22}$$

which is why I have adopted the current conventions. In particular

$$\sigma^{\mu\nu\dagger} = \gamma^0 \sigma^{\mu\nu} \gamma^0$$

$$\text{and since} \quad S = 1 - \frac{i}{4}\omega_{\mu\nu}\sigma^{\mu\nu} + \ldots = \exp\left(-\frac{i}{4}\omega_{\mu\nu}\sigma^{\mu\nu}\right)$$

$$\text{then} \quad S^\dagger = \gamma^0 S^{-1} \gamma^0$$

by inserting lots of $1 = \gamma^0\gamma^0$ in between terms of the expansion. Another way of saying this is

$$\overline{S} \equiv \gamma^0 S^\dagger \gamma^0 = S^{-1} \tag{4.23}$$

and we note that S *is unitary only if* $\omega_{0i} \equiv 0$ leaving only the rotations described by ω_{ij}. (This is no surprise; we noted the pseudo-orthogonality before. This is a finite dimensional representation of a noncompact group.)

It is now obvious how to construct covariant quantities from $\Psi^\dagger$ and we also now see why we put indices on the constant γ^μ matrices. We introduce the adjoint spinor (contravariant) $\overline{\Psi}$ by

$$\overline{\Psi}^\alpha = \Psi^{\dagger\beta}(\gamma^0)^\alpha_\beta \quad \textbf{or} \quad \overline{\Psi} = \Psi^\dagger \gamma^0. \tag{4.24}$$

Then when

$$\Psi' = S\Psi$$

$$\Psi'^\dagger = \Psi^\dagger S^\dagger$$

$$\text{but} \quad \overline{\Psi}' = \overline{\Psi S} = \overline{\Psi} S^{-1}$$

$$\text{that is,} \quad \overline{\Psi}'^\alpha = \overline{\Psi}^\beta (\overline{S})^\beta_\alpha = \overline{\Psi}^\beta (S^{-1})^\alpha_\beta. \tag{4.25}$$

Now

$$S\gamma^\mu S^{-1} = \gamma^\nu \Omega^\mu_\nu, \text{ so}$$

$$\Omega^\mu_\nu S\gamma^\nu S^{-1} = \gamma^\mu$$

$$\therefore S^{-1}\gamma^\mu S = \Omega^\mu_\nu \gamma^\nu \tag{4.26}$$

and hence $\overline{\Psi}\gamma^\mu\Psi$ is a contravariant four-vector.

This extends under space-reflection

$$\underline{x} \to (-)\underline{x}$$
$$t \to t \quad \text{that is} \quad x^{\mu} \to \Sigma_{\mu} g^{\mu\mu} x^{\mu}, \tag{4.27}$$

which you recall has to be adjoined to our proper Lorentz transformations, and we have

$$\Psi \to \Gamma\Psi \text{ with } \overline{\Psi} \to \overline{\Psi P} = \overline{\Psi} P^{-1} \tag{4.28}$$

exactly as earlier. This time $\Gamma^{\mu} = \Sigma_{\mu} g^{\mu\mu} \gamma^{\mu} = \gamma^{\mu} = P^{-1}\gamma^{\mu}P$, which has the (unitary) solution $P = e^{i\phi}\gamma^0$ and if $P^2 = 1$ then $e^{i\phi} = \pm 1$ so usually

$$P = \gamma^0 \tag{4.29}$$

is the standard choice.

We find that all the Γ^R give us $\overline{\Psi}\Gamma^R\Psi$ to be real (Hermitian later when operators), with

$\overline{\Psi}\Psi$	scalar
$\overline{\Psi}\gamma^5\Psi$	pseudoscalar
$\overline{\Psi}\gamma^{\mu}\Psi$	vector
$\overline{\Psi}\sigma^{\mu 5}\Psi$	pseudovector
$\overline{\Psi}\sigma^{\mu\nu}\Psi$	tensor

The Nonrelativistic Limit

Suppose we go to the rest frame of the electron, so that

$$i\frac{\partial\Psi}{\partial t} = H\Psi \to m\gamma^0\Psi \tag{4.30}$$

then we identify

$$\Psi = e^{-imt}\xi \quad \gamma^0\xi = \xi \tag{4.31}$$

$$\Psi = e^{imt}\eta \quad \gamma^0\eta = (-)\eta \tag{4.32}$$

as two sets of solutions. Notice that $\frac{1+\gamma^0}{2}$ is a projection operator for the positive energy solutions and $\frac{1-\gamma^0}{2}$ for the negative.

$$\begin{aligned} P_+ + P_- &= 1 \\ P_+ P_- &= P_- P_+ = 0 \\ P_+ P_+ &= P_+ \\ P_- P_- &= P_-. \end{aligned}$$

There is a representation of the Dirac matrices where

$$\beta = \tau_3 \times 1 = \begin{pmatrix} 1 & 0 \\ 0 & -1 \end{pmatrix} \tag{4.33}$$

$$\underline{\alpha} = \begin{pmatrix} 0 & \underline{\sigma} \\ \underline{\sigma} & 0 \end{pmatrix} = \tau_1 \times \underline{\sigma} \tag{4.34}$$

so that

$$\gamma^0 = \beta = \begin{pmatrix} 1 & 0 \\ 0 & -1 \end{pmatrix} \tag{4.35}$$

$$\gamma^i = \beta\alpha^i = i\tau_2 \times \sigma^i = \begin{pmatrix} 0 & \sigma_i \\ -\sigma^i & 0 \end{pmatrix}. \tag{4.36}$$

In this representation, $(\frac{1+\gamma^0}{2}) = (\begin{smallmatrix} 1 & 0 \\ 0 & 0 \end{smallmatrix})$ and $(\frac{1-\gamma^0}{2}) = (\begin{smallmatrix} 0 & 0 \\ 0 & 1 \end{smallmatrix})$ so that $\Psi = (\begin{smallmatrix} \xi \\ \eta \end{smallmatrix})$. We refer to upper and lower components.

Now in the rest frame, $H = m\gamma^0$ commutes with

$$\Sigma^i = \frac{1}{2}\varepsilon^{ijk}\sigma^{jk} = \begin{pmatrix} \sigma_i & 0 \\ 0 & \sigma_i \end{pmatrix} \tag{4.37}$$

therefore so do $(\frac{1\pm\gamma_0}{2})$. So we can further clarify the solutions by using the projection operators.

Poincaré Group: Inhomogeneous Lorentz Group

Restrict your attention to [4] of space and time. Write

$$x^\mu = (x^0, x^1, x^2, x^3) = (t, x, y, z) = (t, \underline{x}) = (t, x^i) \tag{4.38}$$

where t is really ct but c is set equal to unity. The Lorentz transformations are characterized by

$$\sum_{\mu\nu} \tag{4.39}$$

leaving invariant the quantity

$$dt^2 - dx^2 - dy^2 - dz^2 = g_{\mu\nu} dx^\mu dx^\nu \tag{4.40}$$

$$\text{where} \quad \left. \begin{array}{rcl} g_{00} & = & 1 \\ g_{ij} & = & (-)\delta_{ij} \\ g_{0i} & = & 0 \end{array} \right\} . \tag{4.41}$$

(Note: The relativists, who need $g_{\mu\nu}$ for a metric in curved space, use $\eta_{\mu\nu}$ for this.)

What are the consequences of leaving this second rank covariant metric tensor numerically invariant (form unchanged)? Well you can write this as

$$g_{\mu\nu} = g_{\rho\lambda} \frac{\partial x'^\rho}{\partial x^\mu} \frac{\partial x'^\lambda}{\partial x'^\nu} \tag{4.42}$$

and if we differentiate with respect to $x^\pi \Rightarrow$

$$0 = g_{\rho\lambda} \left\{ \frac{\partial^2 x'^\rho}{\partial x^\mu \partial x^\pi} \frac{\partial x'^\lambda}{\partial x^\nu} + \frac{\partial x'^\rho}{\partial x^\mu} \frac{\partial^2 x'^\lambda}{\partial x^\nu \partial x^\pi} \right\}. \tag{4.43}$$

We rewrite this to emphasize the symmetry as

$$0 = g_{\rho\lambda} \{ S^\rho_{\mu\pi} D^{-1\lambda}_\nu + D^{-1\rho}_\mu S^\lambda_{\nu\pi} \} \tag{4.44}$$

then switching $\mu \leftrightarrow \pi$ and adding, and switching $\pi \leftrightarrow \nu$ and subtracting $\Rightarrow$

$$0 = g_{\rho\lambda} \{ S^\rho_{\mu\pi} D^{-1\lambda}_\nu + \not{D}^{-1\rho}_\mu \not{S}^\lambda_{\nu\pi} + S^\rho_{\pi\mu} D^{-1\lambda}_\nu + \not{D}^\rho_\pi \not{S}^\lambda_{\mu\pi} - \not{S}^\rho_{\mu\nu} \not{D}^{-1\lambda}_\pi - \not{D}^{-1\rho}_\mu \not{S}^\lambda_{\pi\nu} \} \tag{4.45}$$

where the cancellations use the symmetry of both $g_{\rho\lambda}$ and $S^\rho_{\mu\pi}$. We now have

$$0 = 2 g_{\rho\lambda} S^\rho_{\mu\pi} D^{-1\lambda}_\nu \tag{4.46}$$

but both $g_\rho\lambda$ and $D^{-1\lambda}_\nu$ are supposed to be nonsingular, so

$$0 = S^\rho_{\mu\pi} = \frac{\partial^2 x'^\rho}{\partial x^\mu \partial x^\pi}. \tag{4.47}$$

This is solved by having x' be "linear and constant" so that

$$x'^\mu = \Omega^\mu_\nu x^\nu + a^\mu \tag{4.48}$$

where Ω^μ_ν and a^ν are constants. The constant metric condition now reads

$$g_{\mu\nu} = g_{\rho\lambda}\Omega^\rho_\mu\Omega^\lambda_\nu. \tag{4.49}$$

These transformations form the *inhomogeneous Lorentz group* or *Poincaré group*.

We now look at the subgroups of this group. The a^μ parts give translations; they form an abelian subgroup.

Homogeneous (Later Restricted) Lorentz Group

Dropping the translation terms we have

$$x'^\mu = \Omega^\mu_\nu x^\nu \tag{4.50}$$

where the Ω^μ_ν are just numbers, and real ones at that. Notice that the x^μ themselves now form a contravariant vector. We had

$$B^i \to B^j(D^{-1})^i_j, \tag{4.51}$$

so that we identify

$$\Omega^\mu_\nu \equiv (D^{-1})^\mu_\nu. \tag{4.52}$$

Recall that we can raise and lower indices with the numerically invariant metric tensors $g_{\mu\nu}$ and $g^{\mu\nu}$ and notice that $g^{\mu\nu}$ has the same numerical entries as $g_{\mu\nu}$. Notice that

$$\delta^\alpha_\mu = g_{\mu\nu}g^{\nu\alpha} \tag{4.53}$$

$$= \left(g_{\rho\lambda}\Omega^\rho_\mu\Omega^\lambda_\nu\right)g^{\nu\alpha} \quad \text{by invariance of } g_{\mu\nu} \tag{4.54}$$

$$= \Omega^\rho_\mu g_{\rho\lambda}\Omega^\lambda_\nu g^{\nu\alpha} \quad \text{(no change; just a suggestive rewrite)} \tag{4.55}$$

$$= \Omega^\rho_\mu\Omega^\alpha_\rho \tag{4.56}$$

where we have raised and lowered indices on the final symbol. (Note: The position of these indices (as first or last) now has to stay fixed.) But this can be rewritten

$$\delta^\alpha_\mu \equiv (D^{-1})^\rho_\mu\Omega^\alpha_\rho \tag{4.57}$$

so we identify

$$D^\alpha_\rho = \Omega^\alpha_\rho. \tag{4.58}$$

The equation that expresses the fact that $g_{\mu\nu}$ is from invariant, which can also be viewed as a restriction on the allowed elements of Ω or D, can be rewritten in many ways:

$$g_{\mu\nu} = g_{\rho\lambda}\Omega^{\rho}_{\mu}\Omega^{\lambda}_{\nu} \tag{4.59}$$

$$\text{or} \quad \Omega^{\mu}_{\nu}\Omega^{\alpha}_{\mu} = \delta^{\alpha}_{\nu} \tag{4.60}$$

$$\text{or} \quad \Omega^{\mu\nu}\Omega^{\alpha}_{\mu} = g^{\nu\alpha} \tag{4.61}$$

$$\text{or} \quad \Omega^{\mu}_{\nu}g_{\mu\beta}\Omega^{\beta}_{\alpha} = g_{\nu\alpha} \tag{4.62}$$

expressing that it is pseudo-orthogonal. Note that (usually) when authors specify a matrix form they mean

$$x'^{\mu} = \Omega^{\mu}_{\nu}x^{\nu} \tag{4.63}$$

$$\text{as} \begin{bmatrix} x'^0 \\ x'^1 \\ x'^2 \\ x'^3 \end{bmatrix} = \begin{bmatrix} \text{Matrix} \\ \text{Specified} \end{bmatrix} \begin{bmatrix} x^0 \\ x' \\ x^2 \\ x^3 \end{bmatrix} \tag{4.64}$$

and if you slip up on this point, then you will get the wrong signs.

Now consider the last form above as the matrix equation

$$\Omega^T g\Omega = g \tag{4.65}$$

and take the determinant:

$$(\text{ Det } \Omega)^2 = 1 \tag{4.66}$$

$$\therefore \text{ Det } \Omega = \pm 1. \tag{4.67}$$

If Det $\Omega = 1$, we have proper Lorentz transformations and if Det $\Omega = -1$, we have an improper one.

Again, look at the (0, 0) component of the same form:

$$\Omega^{\nu}_{0}g_{\mu\nu}\Omega^{\nu}_{0} = 1, \tag{4.68}$$

that is,

$$\left(\Omega^{0}_{0}\right)^2 - \Sigma_i\left(\Omega^{i}_{0}\right)^2 = 1. \tag{4.69}$$

Hence

$$\left(\Omega^{0}_{0}\right)^2 \geq 1. \tag{4.70}$$

If $\Omega^0_0 \geq 1$, we have an *orthochronous* transformation, and if $\Omega^0_0 \leq (-)1$, we have a *nonorthochronous* one.

There are four sectors:

1. $\mathcal{L}^{\uparrow}_{+}$ with $\Omega^0_0 \geq 1$ and det $\Omega = 1$. Proper, orthochronous subset. These transformations include 1, and clearly form a *subgroup*. We call this

the *restricted homogeneous Lorentz group*. It is characterized by six parameters—three for rotations and three for boosts.

2. $\mathcal{L}_{-}^{\uparrow}$ with $\Omega_0^0 \geq 1$ and det $\Omega = -1$. Improper, orthochronous subset. Not a subgroup. Every Ω in this set can be written as a product of one in $\mathcal{L}_{+}^{\uparrow}$ with the *space inversion*.

$$\Omega(s) = \begin{pmatrix} 1 & & & \\ & -1 & & \\ & & -1 & \\ & & & -1 \end{pmatrix}, \quad \text{that is,} \quad \begin{Bmatrix} t & \rightarrow & t \\ \underline{x} & \rightarrow & (-)\underline{x} \end{Bmatrix}. \tag{4.71}$$

3. $\mathcal{L}_{-}^{\downarrow}$ with $\Omega_0^0 \leq (-)1$ and det $\Omega = -1$. Improper and nonorthochronous. Not a subgroup. Every Ω in this set can be written as a product of one in $\mathcal{L}_{+}^{\uparrow}$ with the *time inversion*.

$$\Omega(t) = \begin{pmatrix} -1 & & & \\ & 1 & & \\ & & 1 & \\ & & & 1 \end{pmatrix}, \quad \text{that is,} \quad \begin{Bmatrix} t & \rightarrow & (-)t \\ \underline{x} & \rightarrow & \underline{x} \end{Bmatrix}. \tag{4.72}$$

4. $\mathcal{L}_{+}^{\downarrow}$ with $\Omega_0^0 \leq 1$ and det $\Omega = 1$. Proper, nonorthochronous. Not a subgroup. Every Ω in this set can be written as a product of one in $\mathcal{L}_{+}^{\uparrow}$ with the *space–time inversion*.

$$\Omega(st) = \Omega(s)\Omega(t) \equiv (-)1 = \begin{pmatrix} -1 & & & \\ & -1 & & \\ & & -1 & \\ & & & -1 \end{pmatrix} \text{ i.e., } \begin{Bmatrix} t & \rightarrow & (-)t \\ \underline{x} & \rightarrow & (-)\underline{x} \end{Bmatrix} \tag{4.73}$$

Notes

1. We usually work with $\mathcal{L}_{+}^{\uparrow}$ and handle the reflections separately.
2. Vectors have an invariant length under $\mathcal{L}$. We call a vector
   time-like if $x^2 > 0$ (e.g., energy and momentum of a free massive particle)
   space-like if $x^2 < 0$ (e.g., momentum transfer)
   null or light-like if $x^2 = 0$ (e.g., energy-momentum of a massless photon)
3. Under $\mathcal{L}_{+}^{\uparrow}$ the sign of the time component is invariant as well as the length of the vector. Then both $\theta(k_0)$ and $\delta(k^2 - m^2)$ are invariants.

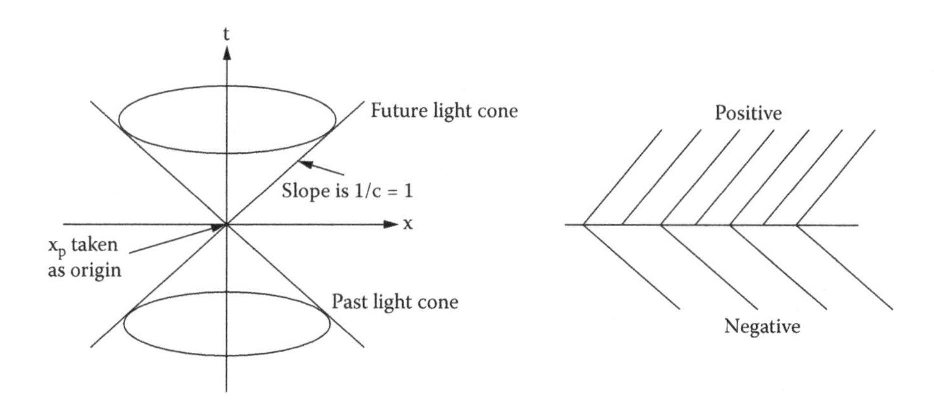

FIGURE 4.2
The light cone of P.

The invariant volume-element in momentum space is

$$\frac{d^4k}{(2\pi)^4}\theta(k_0)\delta(k^2-m^2)2\pi = \frac{d^4k}{(2\pi)^4}\theta(k_0)\delta\left(\left[k_0+\sqrt{k^2+m^2}\right]\left[k_0-\sqrt{k^2+m^2}\right]\right) \tag{4.74}$$

$$= \frac{d^3k}{(2\pi)^3}\frac{1}{2E_k} \quad \text{where} \quad E_k = +\sqrt{\underline{k}^2+m^2} \tag{4.75}$$

Notice the $2E_k$ particles per unit volume (a relativistically covariant statement) and remember to have $(\frac{dk}{2\pi})$ and $2\pi\delta$ to get the factors correct.

4. *Light cone of* P is the set of points such that $(x-xp)^2 = 0$. See Figure 4.2.
5. Obviously, with the structure of A-D, a Lorentz transformation is orthochronous if and only if it transforms every positive time-like vector into another such.
6. Watch the signs!

$$p^\mu = (p^0, p^1, p^2, p^3) = (E, \underline{p}) \tag{4.76}$$

$$p_\mu = (E, (-)\underline{p}) = (p_0, p_1, \ldots) = (p^0, (-)p^1, \ldots) \tag{4.77}$$

(Authors vary.)

$$p^2 = p_\mu p^\mu = p^\mu p^\nu g_{\mu\nu} = E^2 - \underline{p}^2 (= m^2) \tag{4.78}$$

$x^\mu = (t, \underline{x})$ and $x_\mu = (t, -\underline{x})$ so that

$$p \cdot x = p^\mu x_\mu = Et - \underline{p}\cdot\underline{x} \tag{4.79}$$

$$\partial_\mu = \frac{\partial}{\partial x^\mu} = \left(\frac{\partial}{\partial t}, +\underline{\nabla}\right) \text{ but it is covariant} \tag{4.80}$$

while

$$\partial^\mu = \left(\frac{\partial}{\partial t}, -\underline{\nabla}\right) \text{ but it is contravariant.} \tag{4.81}$$

(Really we are mixing vectors and forms.)
Then

$$\partial_\mu V^\mu = \frac{\partial V^0}{\partial t} + \frac{\partial V^x}{\partial x} + = \frac{\partial V^0}{\partial t} + \underline{\nabla} \cdot \underline{V} \tag{4.82}$$

however, the D'Alembertian is

$$\Box = \partial_\mu \partial^\mu = \frac{\partial^2}{\partial t^2} - \nabla^2. \tag{4.83}$$

In the Schrödinger representation, with $\hbar = 1 = c$

$$E \to i\frac{\partial}{\partial t} \tag{4.84}$$

$$\underline{p} \to (-)i\underline{\nabla} \tag{4.85}$$

$$\therefore p_\mu = i\partial_\mu = i\frac{\partial}{\partial x^\mu} \tag{4.86}$$

$$\text{and } p^\mu - i\partial^\mu = i\frac{\partial}{\partial x_\mu}. \tag{4.87}$$

7. Rotation subgroup

$$\Omega^\mu_\lambda \to \begin{pmatrix} 1 & 0 & 0 & 0 \\ 0 & & & \\ 0 & & M & \\ 0 & & & \end{pmatrix} \quad \text{with} \quad MM^T = 1.$$

8. Boosts

$$dx'^\alpha = \Omega^\alpha_\beta dx^\beta. \tag{4.88}$$

Here $d\underline{x} = 0$ and dx' and dy' are zero so

$$dz' = \Omega^3_0 dt \tag{4.89}$$

$$\text{and} \quad dt' = \Omega^0_0 dt. \tag{4.90}$$

$$\text{Now} \quad V = \frac{dz'}{dt'} \Rightarrow \Omega^3 = v\Omega^0_0. \tag{4.91}$$

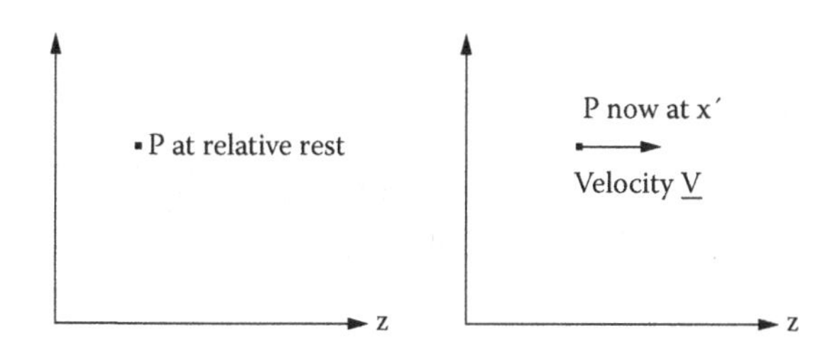

FIGURE 4.3
P at relative rest. P now at x'.

But recall that

$$(\Omega^0_0)^2 - (\Omega^3_0)^2 = 1$$

$$\Omega^0_0 \geq 1 \text{ for } \mathcal{L}^{\uparrow}_{+}.$$

The solution is

$$\Omega^0_0 = \cosh\theta \qquad \Omega^3_0 = \sinh\theta \tag{4.92}$$

$$\text{with} \quad v = \tanh\theta. \tag{4.93}$$

$$\text{that is,} \quad z' = z\cosh\theta + t\sinh\theta \tag{4.94}$$

$$t' = t\cosh\theta + z\sinh\theta. \tag{4.95}$$

Often we call

$$\beta = v\left(= \frac{v}{c}, \text{ of course.}\right) \tag{4.96}$$

$$\gamma = \sqrt{\frac{1}{1-\beta^2}} \tag{4.97}$$

so that

$$z' = \gamma(z + \beta t)$$

$$t' = \gamma(t + \beta z). \tag{4.98}$$

9. A moment's thought should convince you that a general Ω in $\mathcal{L}^{\uparrow}_{+}$ can be written in the form

$$\Omega = \Omega(R_B)\Omega(L_3)\Omega(R_A). \tag{4.99}$$

(a) Rotate the z-axes parallel.
(b) Boost to the desired velocity magnitude.
(c) Rotate to final direction.

Note: There is always ambiguity, and conventions are needed.

The Poincaré Algebra

Recall that $x'^{\mu} = \Omega^{\mu}_{\nu} x^{\nu} + a^{\mu}$ and work with the infinitesimal form

$$\Omega^{\mu}_{\nu} = \delta^{\mu}_{\nu} + \omega^{\mu}_{\nu}$$
$$a^{\mu} = \varepsilon^{\mu} \tag{4.100}$$

so that the constraint equation

$$g_{\mu\nu} = g_{\varrho\lambda}\Omega^{\rho}_{\mu}\Omega^{\lambda}_{\nu}$$

yields

$$\begin{aligned} g^{\mu}_{\nu} &= g_{\varrho\lambda}[\delta^{\rho}_{\mu} + \omega^{\rho}_{\mu}][\delta^{\lambda}_{\nu} + \omega^{\lambda}_{\nu}] \\ &= g_{\mu\nu} + g_{\rho\nu}\omega^{\rho}_{\mu} + g_{\mu\lambda}\omega^{\lambda}_{\nu} + \dots \end{aligned}$$

and thus

$$\omega_{\mu\nu} = (-)\omega_{\nu\mu} \tag{4.101}$$

follows at once. At this stage it is easy to see that ε^{μ} and $\omega_{\mu\nu}$ give you 10 real parameters.

Now expand an element of the group as

$$g = 1 - \frac{i}{2}\omega_{\alpha\beta}M^{\alpha\beta} + i\varepsilon_{\alpha}P^{\alpha} \tag{4.102}$$

where $M^{\alpha\beta}$ is antisymmetric. The operators $M^{\alpha\beta}$ and P^{α} are generators. (Notice the sign in the last term. If we wish to view $\varepsilon^{\alpha} = (\varepsilon^0, \underline{\varepsilon})$ as the shifts, then $g \approx 1 + i\varepsilon^0 P^0 - i\underline{\varepsilon}\cdot\underline{P} = 1 - \varepsilon^0\partial_t - \underline{\varepsilon}\cdot\underline{\nabla}$ as required for $\Psi(x) \to g\Psi(x) = \Psi(T^{-1}x)$ as required in the Schrödinger picture.)

Now we have

$$x'^{\mu} = x^{\mu} + \omega^{\mu}_{\nu}x^{\nu} + \varepsilon^{\mu} \tag{4.103}$$
$$= x^{\mu} + \omega_{\alpha\beta}g^{\mu\alpha}x^{\beta} + \varepsilon_{\alpha}g^{\mu\alpha}. \tag{4.104}$$
$$\therefore x'^{\mu} = x^{\mu} + \frac{1}{2}\omega_{\alpha\beta}[g^{\mu\alpha}x^{\beta} - g^{\mu\beta}x^{\alpha}] + \varepsilon_{\alpha}g^{\mu\alpha}, \tag{4.105}$$

by using the $\omega_{\mu\nu}$ antisymmetry. Hence, from this defining representation we identify: apply g to wave function, not to coordinate.

$$P^{\alpha} = i\partial^{\alpha}$$
$$M^{\alpha\beta} = i(x^{\alpha}\partial^{\beta} - x^{\beta}\partial^{\alpha}). \tag{4.106}$$

Note again the remark on signs made previously.

We can then work out the algebra directly:

$$[P^{\mu}, P^{\nu}] = 0. \tag{4.107}$$

Again:

$$\begin{aligned}[M^{\mu\nu}, P^{\rho}] &= [i(x^{\mu}\partial^{\nu} - x^{\nu}\partial^{\mu}), i\partial^{\rho}] \\ &= g^{\mu\rho}\partial^{\nu} - g^{\nu\rho}\partial^{\mu} \\ &= i(g^{\nu\rho}P^{\mu} - g^{\mu\rho}P^{\nu})\end{aligned}$$

and hence

$$[M^{\mu\nu}, P^{\rho}] = i(g^{\nu\rho}P^{\mu} - g^{\mu\rho}P^{\nu})]. \tag{4.108}$$

Finally

$$\begin{aligned}[M^{\mu\nu}, M^{\rho\lambda}] &= (i)(i)[x^{\mu}\partial^{\nu} - x^{\nu}\partial^{\mu}, x^{\rho}\partial^{\lambda} - x^{\lambda}\partial^{\rho}] \\ &= -\{x^{\mu}g^{\nu\rho}\partial^{\lambda} - x^{\mu}g^{\nu\lambda}\partial^{\rho} - x^{\nu}g^{\mu\rho}\partial^{\lambda} + x^{\nu}g^{\mu\lambda}\partial^{\rho} \\ &\qquad - x^{\rho}g^{\mu\lambda}\partial^{\nu} + x^{\rho}g^{\nu\lambda}\partial^{\mu} + x^{\lambda}g^{\rho\mu}\partial^{\nu} - x^{\lambda}g^{\rho\nu}\partial^{\mu}\} \\ &= g^{\mu\rho}(x^{\nu}\partial^{\lambda} - x^{\lambda}\partial^{\nu}) + g^{\nu\rho}(-x^{\mu}\partial^{\lambda} + x^{\lambda}\partial^{\mu}) \\ &\qquad + g^{\nu\lambda}(x^{\mu}\partial^{\rho} - x^{\rho}\partial^{\mu}) + g^{\mu\lambda}(-x^{\nu}\partial^{\rho} + x^{\rho}\partial^{\nu}) \\ &= i\{g^{\mu\rho}M^{\lambda\nu} + g^{\nu\rho}M\mu\lambda + g^{\nu\lambda}M^{\rho\mu} + g^{\mu\lambda}M^{\nu\rho}\}\end{aligned}$$

so that $$[M^{\mu\nu}, M^{\rho\lambda}] = i\{M^{\mu\lambda}g^{\nu\rho} - M^{\mu\rho}g^{\nu\lambda} + M^{\nu\rho}g^{\mu\lambda} - M^{\nu\lambda}g^{\mu\rho}\}. \tag{4.109}$$

These three sets of commutators characterize the *Poincaré algebra*.

The Casimir Operators and the States

I am sure that you already know that the particles are characterized by mass and spin. Our first task is to exhibit these features. We start with the observation that

$$P^2 = P_{\mu}P^{\mu} \tag{4.110}$$

is a *Casimir operator*. Any Casimir for this problem has to be a homogeneous Lorentz scalar (or pseudoscalar) and thus be characterized by having no free indices. Then it must commute with all the P^{μ}, but this is obvious here.

Now what about other things that commute with the P^{μ}? Well, recall that

$$[M^{\mu\nu}, P^{\rho}] = i(g^{\nu\rho}P^{\mu} - g^{\mu\rho}P^{\nu})$$

so that if we define the *Pauli-Lubanski pseudovector* by

$$W^{\mu} = \frac{1}{2}\varepsilon^{\mu\nu\rho\lambda}P_{\nu}M^{\rho\lambda} \tag{4.111}$$

where the Levi-Civita tensor is antisymmetric as before but with

$$\varepsilon^{0123} = (-)1$$
$$\varepsilon_{0123} = 1$$
$$\text{then} \quad [W^{\mu}, P^{\nu}] = 0. \tag{4.112}$$

This follows from the abelian nature of the P^{μ} subalgebra and the antisymmetry of $\varepsilon^{\mu\nu\rho\lambda}$.

Note: The minus sign in the relative numerical values of ε^{0123} and ε_{0123} is inevitable as a result of raising and lowering. You also now have to have

$$\varepsilon_{\mu\nu\rho\lambda}\varepsilon^{\alpha\beta\gamma\delta} = (-)\left\{\begin{matrix} \delta^{\alpha}_{\mu}\delta^{\beta}_{\nu}\delta^{\gamma}_{\rho}\delta^{\delta}_{\lambda} & - & \delta\delta\delta\delta \\ & \cdots & \\ & \cdots & \end{matrix}\right\} \tag{4.113}$$

where the "extra" minus sign reflects the same remark.

Now to return to our task, all we have to do is to make a homogeneous Lorentz scalar by contracting indices. Obviously $W_{\mu}P^{\nu} \equiv 0$ so the remaining choice is $W^2 = W_{\mu}W^{\mu}$ and we adopt this as our second Casimir.

Now physics motivates us to label the states by all four components of the momentum. So the rest of the complete commuting set of observables (CCSO) must be in the little group of P^{μ} and thus among the W^{μ}. (Detailed enumeration shows there is nothing else. Exercise: Try this for $P^{\mu} = (m, 0, 0, 0)$.)

But what are the commutation relations for the W^{μ} components? Well the W^{μ} form a (pseudo)vector by construction so

$$[M^{\mu\nu}, W^{P}] = i[g^{\nu\rho}W^{\mu} - g^{\mu\rho}W^{\nu}] \tag{4.114}$$

simply by comparison with the P^{μ} transformation law. Again

$$\begin{aligned}
[W^{\mu}, W^{\nu}] &= \left[\frac{1}{2}\varepsilon^{\mu\alpha\beta\gamma}P_{\alpha}M_{\beta\gamma}, W^{\nu}\right] \\
&= \frac{1}{2}\varepsilon^{\mu\alpha\beta\gamma}P_{\alpha}\left[M_{\beta\gamma}, W^{\nu}\right] + zero \\
&= \frac{1}{2}\varepsilon^{\mu\alpha\beta\gamma}P_{\alpha}i\left[W_{\beta}\delta^{\nu}_{\gamma} - W_{\gamma}\delta^{\nu}_{\beta}\right] \\
&= \frac{1}{2}i\left(\varepsilon^{\mu\alpha\beta\nu}P_{\alpha}W_{\beta} - \varepsilon^{\mu\alpha\nu\gamma}W_{\gamma}P_{\alpha}\right)
\end{aligned}$$

$$\text{so that} \quad [\mathbf{W}^{\mu}, \mathbf{W}^{\nu}] = \mathbf{i}\varepsilon^{\mu\nu\alpha\beta}\mathbf{P}_{\alpha}\mathbf{W}_{\beta}. \tag{4.115}$$

On the face of it we are now generating an infinite dimensional algebra, but provided that these W^{μ} act always on states of defined momentum we have effective closure. The precise little group formed by the W^{μ} depends upon the momenta. There are two cases of physical interest.

$$P_\mu|p_\mu> = p_\mu|p_\mu> \quad \text{with} \quad p^2 = m^2. \tag{4.116}$$

We go to the rest frame $p_\mu = (m, 0, 0, 0)$ so that the effective commutators are:

$$[W^i, W^j] = i\varepsilon^{ij0k} P_0 W_k \qquad (W^0 \equiv 0) \tag{4.117}$$

$$= im\varepsilon^{0ijk}(-W^k) \tag{4.118}$$

$$= im\varepsilon^{ijk} W^k \tag{4.119}$$

or

$$\left[\frac{W^i}{m}, \frac{W^j}{m}\right] = \varepsilon^{ijk}\frac{W^k}{m}, \tag{4.120}$$

which we identify as an angular momentum algebra; this is called an SU_2 algebra. Now we can define

$$S^i \equiv \frac{W^i}{m}|_{\text{rest frame}}, \tag{4.121}$$

so that

$$W^2 = (-)m^2 S^2, \tag{4.122}$$

and the final projection of states is

$$|m^2, (-)m^2 s(s+1); \underline{p}, s_3> \tag{4.123}$$

as you probably expected.

$$P_\mu|p_\mu> = p_\mu|p_\mu> \quad \text{with} \quad p^2 = 0. \tag{4.124}$$

We go to the frame where $p^\mu = (E, 0, 0, E)$ and from the definition of W^μ, using $W^\mu p_\mu = 0$ we see that

$$W^\mu = (W^3, W^1, W^2, W^3) \tag{4.125}$$

in this frame. The effective commutation relations are then

$$[W^i, W^j] = i\varepsilon^{ij0k} P_0 W_k + i\varepsilon^{ij30} P_3 W_0 \tag{4.126}$$

$$= i\varepsilon^{0ijk} E((-)W^k) + i(-)\varepsilon^{0ij3}((-)E)(W^3) \tag{4.127}$$

$$= iE\varepsilon W^k - iEW^3\varepsilon^{ij3} \tag{4.128}$$

$$= iE[\varepsilon^{ijk} W^k - \varepsilon^{ij3} W^3] \tag{4.129}$$

and written out in more detail these become

$$[W^1, W^2] = 0 \tag{4.130}$$

$$\left[\frac{W^3}{E}, W^2\right] = iW^1 \tag{4.131}$$

$$\left[\frac{W^3}{E}, W^1\right] = \theta i W^2, \tag{4.132}$$

which we recognize as "translations in the (x, y) plane together with rotations about the z-axis" by the identifications

$$W^1 = \mathcal{P}^1 \tag{4.133}$$

$$W^2 = \mathcal{P}^2, \qquad \text{do not confuse with } P^1 \text{ and so forth} \tag{4.134}$$

$$\frac{W^3}{E} = \mathcal{M}^{12} \tag{4.135}$$

so that

$$[\mathcal{P}^1, \mathcal{P}^2 = 0] \tag{4.136}$$

$$[\mathcal{M}^{12}, \mathcal{P}^1] = (-)i\mathcal{P}^2 \tag{4.137}$$

$$[\mathcal{M}^{12}, \mathcal{P}^2] = i\mathcal{P}^1 \tag{4.138}$$

compares directly with our standard form for "Poincaré." This is called the *Euclidean group in [2].*

There are two quantum numbers to be assigned. We all wave our hands at this point and really appeal to the physics we observe. You set the eigenvalues of $\mathcal{P}^1$ and $\mathcal{P}^2$ to zero (you might have expected continuous labels at this point), which leaves $\mathcal{M}^{12}$ free to have an eigenvalue whereas this would not have been the case. With this in mind you note effectively now

$$W^\mu = \lambda P^\mu \tag{4.139}$$

which is a covariant equation, since P^μ is a vector and W^μ is a (pseudo)vector, provided that λ is a (pseudo)scalar. We call λ the *helicity*. If n_μ is an arbitrary vector, with $n_\mu P^\mu \neq 0$, then

$$\lambda = \frac{n_\nu W^\mu}{n_\mu P^\mu} = \frac{n_\mu \varepsilon^{\mu\nu\rho\lambda} P_\nu M_{\rho\lambda}}{2n_\mu P^\mu} \tag{4.140}$$

and if we take it along the time direction, then

$$\lambda = \frac{1}{2E}\varepsilon^{jk} P^i M^{jk}(-) \tag{4.141}$$

$$= \frac{1}{|\underline{P}|} P^i \frac{\varepsilon^{ijk} M^{jk}}{2}. \tag{4.142}$$

Thus, defining

$$\boxed{J^i = \frac{1}{2}\varepsilon^{ijk} M^{jk}} \tag{4.143}$$

you have

$$\boxed{\lambda = \frac{\underline{p} \cdot \underline{J}}{|\underline{p}|}} \tag{4.144}$$

and we speak of the component of the angular momentum or spin along the three-momentum.

Now, so far you have added only one quantum number (with a couple of zeros to specify the class of representations). When the reflections are added, you connect opposite helicities and the state are finally labeled

$$|m^2 > = |0, |\lambda| \; ; \; \underline{p}, \lambda > .$$

References

1. P.A.M. Dirac, *Proc. Roy. Soc. (London)* A117 (1928): 610.
2. W. Pauli, *Ann. Inst. Henri Poincaré* 6 (1936): 109.
3. E. Schroedinger, *Ann. Physik* 81 (1926): 109.
4. W. Gordon, *Zets für Phys.* 40 (1926): 47.
5. A. Klein, *Zets für Phys.* 41 (1927): 407.
6. E.W. Condon and G.H. Shortley *The Theory of Atomic Spectra*, 4th ed. Cambridge University Press, Cambridge, 1957.
7. H. Casimir, *Proc. R. Acad. Amstd.* 34 (1931): 84.

Problems

4.1 Work through from Equation (4.1) to Equation (4.5) for yourself.

4.2 Work out $\sigma^{\mu 5}\sigma^{\nu 5}$.

4.3 Work out $\sigma^{\mu 5}\gamma^5$.

4.4 Using the Levita-Civita identities, establish $\gamma^\alpha\gamma^\beta\gamma^\gamma = \varepsilon^{\alpha\beta\gamma\mu}\gamma_\mu\gamma^5 + \gamma^\alpha g^{\beta\gamma} - \gamma^\beta g^{\alpha\gamma} + \gamma^\gamma g^{\alpha\beta}$.

4.5 Work out $\gamma^\mu\sigma^{\rho\lambda}$.

4.6 Work out γ^5. Hint: Use the ε form of γ^5 and then $\gamma^\mu\gamma_\mu = 4$.

4.7 Work out $\sigma^{\mu 5}\sigma^{\rho\lambda}$.

4.8 Work out $\gamma^\mu\sigma^{\rho 5}$.

4.9 Work out $\sigma^{\rho\mu}\sigma^{\alpha\beta}$. Hint: Multiply by $\gamma^{\mu}\sigma^{\alpha\beta}$ from the left by γ^{ρ}.

4.10 Establish several matrix representations and check against textbooks.

4.11 Establish several $\sigma \times \tau$ forms where σ and τ are Pauli matrices, for the gamma matrices.

4.12 (For the devoted only.) Try all the previous systematically with s space and t time indices.

4.13 Using the gamma matrix multiplication table show that $\sigma^{\mu\nu}$ do represent the Lorentz algebra.

4.14 Show that $\partial_{\mu} V^{\mu} = \frac{\partial v^0}{\partial t} + \underline{\nabla} \cdot \underline{V}$.

4.15 Show that $\Box \equiv \partial_{\mu}\partial^{\mu} = \frac{\partial^2}{\partial t^2} - \nabla^2$.

4.16 Confirm for yourself that $[M^{\mu\nu}, P^{\rho}] = i(g^{\nu\rho}P^{\mu} - g^{\mu\rho}P^{\nu})$.

4.17 Confirm for yourself that $[M^{\mu\nu}, M^{\rho\lambda}] = i\{M^{\mu\lambda}g^{\nu\rho} - M^{\mu\rho}g^{\nu\lambda} + M^{\nu\rho}g^{\mu\lambda} - M^{\nu\lambda}g^{\mu\rho}\}$.

4.18 Confirm for yourself that $[M^{\mu\nu}, W^{\rho}] = i[g^{\nu\rho}W^{\mu}i - g^{\mu\rho}W^{\nu}]$.

4.19 Confirm for yourself that $[W^{\mu}, W^{\nu}] = i\varepsilon^{\mu\nu\alpha\beta}P_{\alpha}W_{\beta}$.

4.20 Confirm for yourself that if $\underline{s}^2 \equiv \underline{s} \cdot \underline{s} = 1$, where $\underline{s}$ is called a "spin-polarization vector," that $\frac{1}{2}(1+\underline{s}\cdot\underline{\sigma})$ and $\frac{1}{2}(1-\underline{s}\cdot\underline{\sigma})$, where $\underline{\sigma}$ are the Pauli matrices, satisfy that each of their squares is unity but their product in either order is zero. These are called projection operators. They and many of their generalizations are very convenient for doing calculations.

5

The Internal Symmetries

This chapter deals with the internal symmetries of the standard model of elementary particle physics. We start by considering global symmetries and the associated local conservation laws. We have already met some of this structure before in the form of $U(1)$ and $SU(2)$ except that they now act on the fields that create and annihilate the elementary particles rather than on space–time itself. We take a global $U(1)$ represented by

$$U(1) = \exp\left(\frac{-i}{2}\theta N\right) \tag{5.1}$$

where θ is a constant parameter and N is an operator in an internal space where its eigenvalues are the numbers associated with the physical states of the system or the quantum fields that create and annihilate them. This allows quantum numbers (e.g., charges) to be assigned to the states and fields. Now if θ is indeed a constant then $U(1)$ will produce global symmetries, which do not vary from place to place, or time to time, and so forth. The consequence of this is that there are local conservation laws, which force certain labels on the states and fields to either stay constant or to only change such that certain specific combinations are constant. This puts strict conditions on the interactions between the states.

It may well be that you have no trouble in understanding what has just been presented. On the other hand, if you have no previous experience, it may be meaningless to you. With this in mind I offer two simple examples, quite unrelated to the physics, in the hope that they may help.

The first one consists of a puzzle made up from the 62 remaining squares of a familiar chess or draughts board when two diagonally opposite corner squares are removed. You are also given 31 dominoes, each two squares by one in size in terms of the chessboard. The total areas of the mutilated chessboard and the full set of dominoes are equal. The question is whether you can cover the one object with the other without further mutilations of either. To solve this puzzle consider placing a domino over two adjacent squares on the board. Clearly it covers two squares of different colors. Again consider the color of one missing corner of the board and its relationship to the diagonally opposite one. Clearly, they are of the same color. So the task cannot be done. A discrete symmetry has answered a "state of the system" problem.

DOI: 10.1201/9781439895207-5

The next puzzle is quite different. You are given two identical wine glasses, one half full of red wine and the other half full of white wine. You also have a small ladle (or spoon). This puzzle comes in several stages. First take a spoonful from the red wine and transfer it into the white wine. Question: Is the concentration of white in the red glass greater than the concentration of the red in the white glass or are they equal? You should easily see that they are equal. (Try doing this as a piece of algebra; it is quite easy.)

Now the process is repeated except that after the red wine is put into the white wine glass it is stirred with the spoon. Then the process is completed as before. What is your answer now? It is the same. (Try the algebra. It is harder but still possible.) Obviously something "deep" is going on. There is a conserved (unchanging) quantity involved. It is the difference between the amount of white wine transferred overall and the amount of red wine transferred overall in the opposite direction. This is clearly zero. You can work this out by algebra yet again, or observe that the initial and final levels in the two glasses are the same. This time a (continuous) symmetry has ensured that the result is unchanged.

It is probably about time that we looked at the Noether theorem [1].

Suppose that we have a Lagrangian density

$$\mathcal{L} = \mathcal{L}(\psi^\alpha(x), \partial_\mu \psi^\alpha(x)) \tag{5.2}$$

and that we write

$$\frac{\partial \psi^\alpha(x)}{\partial \mu} = \frac{\partial \psi^\alpha(x)}{\partial x^\mu} = \psi^\alpha_\mu(x). \tag{5.3}$$

Then the Lagrangian $L = \int \mathcal{L} d^3x$ is invariant under the changes of fields

$$\delta \psi^\alpha(x) \tag{5.4}$$

and the current density

$$j^\mu(x) = \text{ by definition } = \frac{\partial \mathcal{L}}{\partial \psi^\alpha_\mu} \delta \psi^\alpha \tag{5.5}$$

$$\text{has } \partial_\mu j^\mu \equiv 0. \tag{5.6}$$

The Lie algebra Λ

$$[\chi_\alpha, \chi_\rho] = i f_{\alpha i \rho} \chi_i \tag{5.7}$$

where the f_{ijk} are the constant structure constants, implies that

$$J^\mu_\rho(x) = (-i) \frac{\partial L}{\partial \psi^\alpha_\mu(x)} [\chi_\rho]^\alpha_\beta \psi^\beta(x). \tag{5.8}$$

Invariance of L under Λ tells us that

$$\partial_\mu J^\mu_\rho(x) = 0 \tag{5.9}$$

$$\text{or} \quad \frac{\partial}{\partial t}\int J^\mu_\rho(x) d\,\text{Vol} = 0. \tag{5.10}$$

Armed with this knowledge we now turn to the modern way of viewing forces. We start with the constant metric tensor $\eta^{\mu\nu}$, the components of which are given by

$$\eta^{00} = 1 \tag{5.11}$$

$$\eta^{ij} = \delta^{ij} = 0 \text{ if } i \neq j \quad \text{but} \quad 1 \text{ if } i = j \tag{5.12}$$

and all off diagonal elements are equal to zero. (The Kronecker delta, δ^{ij}, may be familiar.)

There are two very useful properties of δ^{ij}. Clearly $\delta^{ij}A_j = A^i$, which exhibits the raising of indices. The student is urged to check the lowering of indices also with δ_{ij}. Again $\delta^{ij} = 1 + 1 + 1 = 3$, which is the number of spatial indices.

It is convenient at this stage to introduce the properties of the totally antisymmetric Levi-Civita tensor ε_{ijk}. By inspection

$$\begin{aligned}\varepsilon_{ijk}\varepsilon^{lmn} &= \delta_i{}^l\delta_j{}^m\delta_k{}^n - \delta_i{}^l\delta_j{}^n\delta_k{}^m \\ &+ \delta_i{}^m\delta_j{}^n\delta_k{}^l - \delta_i{}^n\delta_j{}^m\delta_k{}^l \\ &+ \delta_i{}^n\delta_j{}^l\delta_k{}^m - \delta_i{}^m\delta_j{}^l\delta_k{}^n.\end{aligned} \tag{5.13}$$

Now do not throw up your hands in horror and despair. Let us pick *ijk* to be 123 and lmn also to be 123. (You must watch the order of indices because ε_{ijk} is antisymmetric.) We note that this checks the sign of the top term in the left-hand column, because $\varepsilon_{123}\varepsilon^{123} = 1 \times 1 = 1$ and $\delta_1^1\delta_2^2\delta_3^3 = 1 \times 1 \times 1 = 1$. Now look down the left-hand column and notice that *ijk* are fixed but lmn are cyclic, so that confirms the signs of the two lower entries. Again look at the top of the right-hand column and notice that *ijk* are again fixed but one switch of a pair of indices ($n \leftrightarrow m$) has been made, so that confirms the minus sign. Finally, look at the two lower members of the right-hand column to notice that *ijk* are again fixed but the lnm are now just cycled, confirming the signs. This is the way to remember the full original formula. Do not just trust your memory. Use the symmetries and antisymmetries. Now consider $\varepsilon_{ijk}\varepsilon^{lmk}$, which is just the original expression we started from, but n has been picked equal to k and is therefore now summed, leaving us with a reduced tensor with only two upper and two lower indices. It is simple to work this out directly:

$$\begin{aligned}\varepsilon_{ijk}\varepsilon^{lmk} &= \delta_i{}^l\delta_j{}^m\delta_k{}^k - \delta_i{}^l\delta_j{}^k\delta_k{}^m \\ &\quad + \delta_i{}^m\delta_j{}^k\delta_k{}^l - \delta_i{}^k\delta_j{}^m\delta_k{}^l \\ &\quad + \delta_i{}^k\delta_j{}^l\delta_k{}^m - \delta_i{}^m\delta_j{}^l\delta_k{}^k \\ &= \delta_i{}^l\delta_j{}^m 3 - \delta_i{}^l\delta_j{}^n \\ &\quad + \delta_i{}^m\delta_j{}^l - \delta_i{}^l\delta_j{}^m \\ &\quad + \delta_i{}^m\delta_j{}^l - \delta_i{}^m\delta_j{}^l 3 \\ &= \delta_i{}^l\delta_j{}^m - \delta_i{}^m\delta_j{}^l. \end{aligned} \tag{5.14}$$

Again consider

$$\begin{aligned}\varepsilon_{ijk}\varepsilon^{ljk} &= \delta_i{}^l\delta_j{}^j - \delta_i{}^j\delta_j{}^l \\ &= 3\delta_i{}^l - \delta_i^l \\ &= 2\delta_i{}^l. \end{aligned} \tag{5.15}$$

Finally consider

$$\varepsilon_{ijk}\varepsilon^{ijk} = 2\delta_i{}^i = 3!. \tag{5.16}$$

It always works to the pattern shown, and the final coefficient (here 3!) is always the factorial of the dimension of the case considered with previous ones going back up the list being the (dimension - 1)! and so forth as the tensor surviving has more indices. Keen students may try to work out the case for three spatial indices and one time index (the world we live in!) carefully watching the signs.

Armed with this mathematics we turn to studying electromagnetism and local gauge invariance. Suppose that charge conservation has been imposed on the coupling of a photon (the electromagnetic field potential) for a charged particle. In terms of the 4-potential

$$A^\mu(x) = (\Phi(x), A^i(x)) = g^{\mu\nu}A_\nu(x) \tag{5.17}$$

where $\Phi(x)$ is the scalar potential and the $A^i(x)$ are the three components of the vector potential. The field strengths are defined by

$$F^{\mu\nu} = \frac{\partial}{\partial x_\nu}A^\mu - \frac{\partial}{\partial x_\mu}A^\nu \tag{5.18}$$

and the electric field and magnetic induction in a noncovariant notation are given by

$$\underline{E} = (F^{01}, F^{02}, F^{03}) \text{ and } \underline{B} = (F^{23}, F^{31}, F^{12}), \tag{5.19}$$

respectively. You are urged to check this out for yourself, taking particular care with the signs.

The free Dirac equation of mass m reads

$$(i\,\hbar\,\not{\nabla} - mc)\psi = 0 \tag{5.20}$$

$$\text{where}\quad \not{\nabla} = \gamma^{\mu}\frac{\partial}{\partial x^{\mu}} = \frac{\gamma^{0}}{c}\frac{\partial}{\partial t} + \underline{\gamma}\cdot\underline{\nabla} \tag{5.21}$$

where $\hbar$ is Planck's constant divided by 2π and c is the velocity of light. (These constants will be dropped in what follows, bringing us to so-called natural units for the subject.) In momentum space with

$$p^{\mu} = i\,\hbar\,\frac{\partial}{\partial x_{\mu}} \tag{5.22}$$

the Dirac equation takes the form

$$(\not{p} - mc)\psi = 0. \tag{5.23}$$

You may well have met the concept of "minimal" substitution where the interaction of electromagnetism is introduced by

$$\left(\not{p} - \frac{e\,\not{A}}{c} - mc\right)\psi = 0. \tag{5.24}$$

In many ways it is a pity that this choice of sign was made for the charge on the electron (the choice is arbitrary) as it leads to endless confusion in conduction of currents down wires. There is no way anyone can change this after so many years of history—certainly not this author.

From a modern viewpoint the coupling given follows from local gauge invariance, or gauge invariance of the second kind as it is sometimes called. However, we will start from global gauge invariance or gauge invariance of the first kind. Here there is for $U(1)$ a parameter θ, which depends on neither space nor time. The electron transforms as

$$\psi \to \exp\left(\frac{-i\theta}{2}\right)\psi \tag{5.25}$$

and the electromagnetic four potential transforms as

$$A^{\mu} \to \exp\left(\frac{-i}{2}\theta\right)A^{\mu}\exp\left(\frac{i}{2}\theta\right), \tag{5.26}$$

so that taking the derivative form of the coupled Dirac equation

$$(i\not{\partial} - \frac{e}{c}\not{A} - mc)\psi = 0 \tag{5.27}$$

and we see that it is unchanged. Curiously, perhaps, there is already information to be found here. Since electric charge is conserved, and the space and time dependence of the fields is local, then creation and annihilation of charge must also simultaneously be local. However, this is not what we are looking for. Instead, let θ depend on both space and time, so that the invariances are now local. This time the derivative in the derivative form of the coupled Dirac equation (Equation 5.27) seems to spoil the symmetry when it acts on the parameter θ. However, the local transformation of the four potential is then assumed to change and thus restore the symmetry. We take

$$\psi(x) \rightarrow e^{\frac{-i\theta(x)}{2}} \psi(x) \tag{5.28}$$

where we see that because of the x dependence of θ we must now look to restore a local symmetry. If we can find this, we would be looking at abelian gauge invariance, because there is no nontrivial group theory involved. We take a Lagrangian density for the Dirac electron field interacting with the electromagnetic current of the form

$$\mathcal{L} = \overline{\psi}(i\partial\!\!\!/ - m)\psi - \frac{1}{4}F_{\mu}F^{\mu} - e\overline{\psi}A\!\!\!/\psi \tag{5.29}$$

in the natural units of the subject where $\hbar = 1 = c$. Here

$$F^{\mu\nu} = \partial^{\mu} A^{\nu} - \partial^{\nu} A^{\mu} \tag{5.30}$$

and we require the vector field to transform as

$$A_{\mu} \rightarrow A_{\mu} - i\partial_{\mu}\frac{\theta(x)}{2} \tag{5.31}$$

so that $F^{\mu\nu}$ is invariant and then define a covariant derivative by

$$D_{\mu}\psi = \left(\partial_{\mu} + i\frac{e}{2}A_{\mu}\right)\psi, \tag{5.32}$$

so that we see that $D_{\mu}\psi$ transforms in the same way as ψ

$$D_{\mu}\psi \rightarrow e^{\frac{-ie}{2}\theta(x)} D_{\mu}\psi \tag{5.33}$$

and our Lagrange density is invariant with the conventional factor of $\frac{1}{4}$ as stated. This describes a massless vector field (experimentally the photon), because the mass term proportional to $A_{\mu}A^{\mu}$ is not invariant.

We now move onto consider nonabelian gauge field theories. The extension required is to write

$$\psi(x) \rightarrow e^{-ig\frac{\theta_i}{2}T_i} \psi(x) \tag{5.34}$$

where the T_i are square matrices satisfying the commutation relations of some nonabelian Lie algebra

$$[T_i, T_j] \equiv T_i T_j - T_j T_i = i f_{ijk} T_k \tag{5.35}$$

where the f_{ijk} are the totally antisymmetric structure constants of the Lie group. For $SU(2)$ these are simply the ε_{ijk} Levi-Civita tensor components. By analogy with the previous case we introduce a covariant derivative by

$$D^\mu \psi = \left(\partial^\mu + \frac{ig}{2} \underline{T} \cdot \underline{\theta}(x) \right) \psi \tag{5.36}$$

where g is the coupling constant (rather like e) and $\theta_i(x)$ has been extended to have the same number of components as the adjoint (or regular) representation of the Lie group, for example, 3 for $SU(2)$. Thus, there is now the corresponding number of gauge fields, and we adopt the gauge transformation property

$$A_i^\mu \to A_i^\mu \to i\partial^\mu \frac{\theta_i(x)}{2} + \frac{g}{2} f_{ijk} \theta_j A_k^\mu \tag{5.37}$$

to allow the gauge fields to cancel out the unwanted terms. To construct a gauge invariant Lagrangian for the gauge fields themselves we define

$$F^{\mu\nu} = F_i^{\mu\nu} T_i = -ig^{-1}[D^\mu, D^\nu] \tag{5.38}$$

so that

$$F^{\mu\nu} = \partial^\mu A^\nu - \partial^\nu A^\mu + ig[A^\mu, A^\nu] \tag{5.39}$$

where a total derivative has been dropped.

Equivalently we have

$$F_i^{\mu\nu} = \partial^\mu A_i^\nu - \partial^\nu A_i^\mu - g f_{ijk} A_j^\mu A_k^\nu \tag{5.40}$$

and this is independent of the fermion representation.

The transformation of $F^{\mu\nu}$ under the gauge group is

$$F^{\mu\nu}(x) \to U(x) F^{\mu\nu}(x) U^{-1}(x) \tag{5.41}$$

so that a gauge invariant Lagrangian $\mathcal{L}_{YM}$ for the gauge (now Yang–Mills) [2] fields is

$$\begin{aligned} \mathcal{L}_{YM} &= (-)\frac{1}{2} Tr \cdot (F_{\mu\nu} F^{\mu\nu}) \\ &= (-)\frac{1}{4} F_i^{\mu\nu} F_{\mu\nu}^i \end{aligned} \tag{5.42}$$

since the normalization of the generators of gauge group is conventionally given by

$$Tr\left(\frac{T_i}{2}\frac{T_j}{2}\right) = \frac{1}{2}\delta_{ij}. \tag{5.43}$$

We can see that the gauge invariant Lagrangian for a Dirac spinor field interacting with vector gauge fields is

$$\begin{aligned}\mathcal{L} &= \overline{\psi}(i\not{D} - m)\psi - \frac{1}{2}Tr \cdot (F_{\mu\nu}F^{\mu\nu}) \\ &= \overline{\psi}(i\not{D} - m)\psi - \frac{1}{4}F^i_{\mu\nu}F^{\mu\nu}_i. \end{aligned} \tag{5.44}$$

If the gauge group is a simple Lie group, then there is a single coupling constant g. But, if the gauge group is semisimple, one which can be written as a product of simple factors, then there will be separate coupling constants for each factor.

For our case, the Euler–Lagrange equations are

$$\partial^\nu \frac{\partial \mathcal{L}}{\partial(\partial^\nu A^\mu_i)} = \frac{\partial \mathcal{L}}{\partial A^\mu_i}, \tag{5.45}$$

$$\partial^\nu \frac{\partial \mathcal{L}}{\partial(\partial^\nu \overline{\psi})} = \frac{\partial \mathcal{L}}{\partial \overline{\psi}}. \tag{5.46}$$

The first of these leads to

$$\partial^\mu F^i_{\mu\nu} - g f_{ijk} A^\mu_j F^k_{\mu\nu} = g\overline{\psi}\gamma^\mu T_i \psi, \tag{5.47}$$

which can be rewritten as

$$D^\mu A^\nu_i = \partial^\mu A^\nu_i - g f_{ijk} A^\mu_j A^\nu_k \tag{5.48}$$

in terms of the covariant derivative of the gauge field. The other one gives

$$(i\not{D} - m)\psi = 0 \tag{5.49}$$

$$\text{with} \quad D^\mu \psi = \left(\partial^\mu + \frac{ig}{2}\underline{T} \cdot \underline{\delta}(x)\right)\psi \tag{5.50}$$

as previously.

Because we need $SU(3)$ for a treatment of quantum chromodynamics in the standard model, we shall simply list the main features here. Because the colors of the quarks, which we shall take as red, blue, and green, are three in number, we shall need an extension of the 2×2 Pauli matrices to 3×3 matrices and correspondingly the adjoint or regular representation will be 8×8. We start by defining

$$\lambda_1 = \begin{pmatrix} 0 & 1 & 0 \\ 1 & 0 & 0 \\ 0 & 0 & 0 \end{pmatrix} \quad \lambda_2 = \begin{pmatrix} 0 & -i & 0 \\ i & 0 & 0 \\ 0 & 0 & 0 \end{pmatrix} \quad \lambda_3 = \begin{pmatrix} 1 & 0 & 0 \\ 0 & -1 & 0 \\ 0 & 0 & 0 \end{pmatrix} \tag{5.51}$$

exactly as with Pauli matrices but only in the top left-hand corner. The commutators of pairs of these

$$[\lambda_i, \lambda_j] = 2i f_{ijk} \lambda_k \tag{5.52}$$

reveal that the *structure constants* in this sector are simply the ε_{ijk} of Levi-Cevita. There is a fourth matrix, usually designated λ_8, which commutes to zero with the first three, and with the standard normalization given by

$$Tr \cdot [\lambda_8, \lambda_8] = 2 \tag{5.53}$$

is taken to be

$$\lambda_8 = \frac{1}{\sqrt{3}} \begin{bmatrix} 1 & & \\ & 1 & \\ & & -2 \end{bmatrix}. \tag{5.54}$$

Then λ_4 and λ_5, copying the first two Pauli matrices but in the second and third columns, are taken as

$$\lambda_4 = \frac{1}{\sqrt{2}} \begin{bmatrix} 0 & 0 & 1 \\ 0 & 0 & 0 \\ 0 & 1 & 0 \end{bmatrix} \quad \text{and} \quad \lambda_5 = \frac{1}{\sqrt{2}} \begin{bmatrix} 0 & 0 & -i \\ 0 & 0 & 0 \\ 0 & i & 0 \end{bmatrix}. \tag{5.55}$$

Similarly, λ_6 and λ_7 become

$$\lambda_6 = \frac{1}{\sqrt{2}} \begin{bmatrix} 0 & 0 & 0 \\ 0 & 0 & 1 \\ 0 & 1 & 0 \end{bmatrix} \quad \text{and} \quad \lambda_7 = \begin{bmatrix} 0 & 0 & 0 \\ 0 & 0 & -i \\ 0 & i & 0 \end{bmatrix}, \tag{5.56}$$

making $8 = 3^2 - 1$ in total.

The totally antisymmetric structure constants f_{ijk} are

ijk	f_{ijk}
123	1
147	$\frac{1}{2}$
156	$-\frac{1}{2}$
246	$\frac{1}{2}$
257	$\frac{1}{2}$
345	$\frac{1}{2}$
367	$-\frac{1}{2}$
458	$\frac{1}{2}\sqrt{3}$
678	$\frac{1}{2}\sqrt{3}$

with all nonindicated ones vanishing. However, there are now some totally symmetric constants denoted by d_{ijk}, which are

ijk	d_{ijk}
118	$\frac{1}{\sqrt{3}}$
228	$\frac{1}{\sqrt{3}}$
338	$\frac{1}{\sqrt{3}}$
146	$\frac{1}{2}$
157	$\frac{1}{2}$
247	$-\frac{1}{2}$
256	$\frac{1}{2}$
344	$\frac{1}{2}$
355	$\frac{1}{2}$
366	$-\frac{1}{2}$
377	$-\frac{1}{2}$
448	$-\frac{1}{2\sqrt{3}}$
558	$-\frac{1}{2\sqrt{3}}$
668	$-\frac{1}{2\sqrt{3}}$
778	$-\frac{1}{2\sqrt{3}}$
888	$-\frac{1}{\sqrt{3}}$

with all nonindicated ones vanishing. You can see that there are eight colors of massless vector bosons, called gluons in this case. They couple only to quarks carrying three colors (normally called red, blue, and green) and self-coupling in the now familiar manner. These are the strong interactions and they are believed to be confining so that there is no free color. The general idea is that if we try to separate free color then the forces between them become so strong that new pairs of opposite and equal color changes are created and the free color remains hidden. There is no solution to this nonlinear field system. The usual attack is to move to a four-dimension space (time becomes the fourth space dimension) then to introduce a lattice structure and to attempt to find approximate solutions using very extensive national computer systems. The coupling constant g is found to change with energy scale. There is no known way to unify this strong force with the weak and electromagnetic forces.

References

1. E. Noether, *Nachr. Akad, Wiss. Götingen Math.-Phys. Kl.* II (1918): 235. M.A. Tavel, *Transport Theory Statist. Phys.* 1, no. 3 (1971): 183.
2. C.N. Yang and R.L. Mills, *Phys. Rev.* 96 (1954): 191.

Problems

5.1 In the red and white wine puzzle, do the algebra in the easy case.

5.2 In the red and white wine puzzle, do the algebra in the harder case.

5.3 Without looking back at the text, write down $\varepsilon_{ijk}\,\varepsilon^{lmn}$ and then check your result.

5.4 Again without looking back at the text, find the contracted forms of the product of $\varepsilon_{ijk}\,\varepsilon^{lmn}$ and then check your results.

5.5 Check Equation (5.19) making sure that you get the signs correct.

5.6 Read again from Equations (5.28) to (5.33). Then repeat this calculation until you get it correct.

5.7 Read again the section leading to the covariant derivative in terms of the derivative of the four potential and the structure constants f_{ijk} in Equation (5.48). Repeat without looking back at the text until you can confidently perform this construction.

5.8 Write down the Pauli matrices and the unit 2×2 matrix. Work out all their products.

5.9 Extend the work in Problem 5.8 to the 3×3 case, making sure that you understand the logic.

5.10 Calculate the f_{ijk} structure constants for the 3×3 case and check your results against the text.

6

Lie Group Techniques for the Standard Model Lie Groups

The student who is happy to build a strong background in these topics is advised to consult H. Georgi [1]. Those who prefer a detailed mathematic derivation are advised to consult J.F. Cornwell [2]. The techniques for finding the explicit forms of the characters of the tensor irreducible representations of the unitary groups, symplectic groups, and both sets of orthogonal groups can be found in N.E.P. Samara and R.C. King [3].

The case of unitary groups has already been treated. I now present the cases of the symplectic groups and the orthogonal groups, which are extracted from Samara and King [4] in a manner designed to show the N dependence in each case, as this is most frequently the information really required.

For $SO(N)$ in terms of the partition labels λ we find

$$D_N[\lambda] = \begin{bmatrix} \underset{(i \geq j)}{\overset{\lambda}{\pi}} \quad (N + \lambda_i + \lambda_j - i - j)\chi \\ \underset{(i \geq j)}{\overset{\lambda}{\pi}} \quad (N - \lambda_i^N - \lambda_j^N + i + j - 2) \end{bmatrix} H(\lambda)| > , \tag{6.1}$$

and for $SP(N)$, in a similar way, we find

$$D_N < \lambda > = \begin{bmatrix} \underset{(i > j)}{\overset{\lambda}{\pi}} \quad (N + \lambda_i + \lambda_j - i - j + 2)x \\ \underset{(i \leq j)}{\overset{\lambda}{\pi}} \quad (N - \lambda_i^N - \lambda_j^N + i + j) \end{bmatrix} H(\lambda)| > . \tag{6.2}$$

DOI: 10.1201/9781439895207-6

Roots and Weights

This will probably be familiar to the reader from quantum mechanics in operator form. We want to find a complete commuting set of observables. In this case we write them as a Cartan subalgebra of Hermitian operators:

$$H_i = \delta_i^\dagger > \tag{6.3}$$

$$[H_i, H_j] = 0. \tag{6.4}$$

We can normalize by

$$\text{Tr.}(H_i H_j) = k_D \delta_{ij} \quad \text{for} \quad i, j = 1 \text{ to } m, \tag{6.5}$$

where k_D depends on the representation and the normalization, and m is called the rank of the algebra.

The states of a representation D can be designated by

$$H_i|\mu, x, D> = \mu_i|\mu, x, D> \tag{6.6}$$

after diagonalization. The μ^i are called weights and are real. The m-component vector made of the μ^i is the weight vector. Any other label that is needed to specify the state is denoted by x.

The adjoint equation is particularly important. It has the rows and columns of the matrices labeled by the same index that labels the generators. We can call the state in the adjoint representation corresponding to an arbitrary operator V_i as $|V_i>$. The scalar product is taken as

$$< V_i|V_j> = (k_D)^{-1}\,\text{Tr}(V_i^\dagger V_j)>, \tag{6.7}$$

where the dagger is included to allow for complex linear combinations of generators needed when we raise and lower states in quantum mechanics on operators for $SU(2)$. The action of a generator on a state can be calculated as

$$\begin{aligned} V_i|V_j> &= |V_k>< V_k|V_i|V_j> \\ &= |V_k>[T_i]_{kj} \\ &= -if_{ikj}|V_k> \\ &= if_{ijk}|V_k>, \end{aligned} \tag{6.8}$$

which is, of course, just the set for the commutator of V_i and V_k.

The *roots* are the weights of the adjoint representation. Because of the Cartan generators, commute the corresponding states have zero weight vectors. All states in the adjoint representation with zero weight vectors correspond to Cartan generators. The other states in the adjoint representation have nonzero weight vectors α_i with components α, so that

$$H_i|E_\alpha> = \alpha_i|E_\alpha>, \tag{6.9}$$

which implies that

$$[H_i, E_\alpha] = \alpha_i E_\alpha. \tag{6.10}$$

Notice that the E_α are not Hermitian just like raising and lowering operators in $SU(2)$. The normalization of states in the adjoint representation is taken to ensure

$$< E_\alpha|E_\beta > = \lambda^{-1}\,\mathrm{Tr}(E_\alpha^\dagger E_D) = \delta_{\alpha\beta}. \tag{6.11}$$

The weights α are called roots and the special weight vector α with components α_i is a *root vector.*

You may enjoy showing that

$$[E_\alpha, E_{-\alpha}] = \alpha.H, \tag{6.12}$$

which should remind you of the $SU(2)$ commutation relation $[J^+, J^-] = J_3$. This analogy will be exploited to learn more about the representations of compact Lie groups. Generally, for any weight μ of a representation D, the E_3 value is given by

$$E_3|\mu, x, D> = \frac{\alpha\cdot\mu}{\alpha^2}|\mu, x, D> . \tag{6.13}$$

Because the E_3 values must be integers or half integers,

$$\frac{2\alpha\cdot\mu}{\alpha^2} = \text{integer}. \tag{6.14}$$

It is simple to show that

$$\frac{\alpha\cdot\mu}{\alpha^2} = -\frac{1}{2}(p-q) \tag{6.15}$$

where $\mu + p\alpha$ is the weight of the highest E_3 state of the $SU(2)$ spin j representation and q plays the corresponding role when lowering.

We can easily make use of these results. Apply Equation (6.13) to both distinct roots α and β and we find (using E_α as the $SU(2)$ definition)

$$\frac{\alpha\cdot\beta}{\alpha^2} = (-)\frac{1}{2}(p-q). \tag{6.16}$$

Again, now using E_β yields instead

$$\frac{\beta\cdot\alpha}{\beta^2} = (-)\frac{1}{2}(p'-q'). \tag{6.17}$$

If $\theta_{\alpha\beta}$ is the angle between the roots α and β then multiplying the last two results gives

$$\cos^2(\theta_{\alpha\beta}) = \frac{(\alpha\cdot\beta)^2}{\alpha^2\beta^2} = \frac{(p-q)(p'-q')}{4}. \tag{6.18}$$

Now this is important. Notice that $(p-q)(p'-q')$ must be an integer that is nonnegative so that there exist just four possibilities (up to complements) for the angles and between the roots.

We can list these possibilities:

$(p-q)(p'-q')$	$\theta_{\alpha\beta}$
0	90°
1	60° or 120°
2	45° or 135°
3	30° or 150°

You may think that we have missed two other possibilities. But $(p-q)(p'-q') = 4$ corresponds to 0° or 180°. Neither is of any use. The first is in violation of uniqueness. The second is trivial because roots come in pairs of opposite signs, both in the same $SU(2)$ group.

We have met $SU(3)$ before, but it will be convenient to learn a little more. What are the weights and the roots? Well T_3 and T_8 are already diagonal and normalized in the standard way. The eigenvectors and associated weights are

$$\begin{pmatrix}1\\0\\0\end{pmatrix} \to \left(\frac{1}{2}, \frac{\sqrt{3}}{6}\right) \begin{pmatrix}0\\1\\0\end{pmatrix} \to \left(-\frac{1}{2}, \frac{\sqrt{3}}{6}\right) \begin{pmatrix}0\\0\\1\end{pmatrix} \to \left(0, -\frac{\sqrt{3}}{3}\right). \tag{6.19}$$

These vectors, plotted in a plane, form the vertices of an equilateral triangle (Figure 6.1).

The roots are differences of weights, because the corresponding generators take one weight to another. The generators clearly have only one-off diagonal

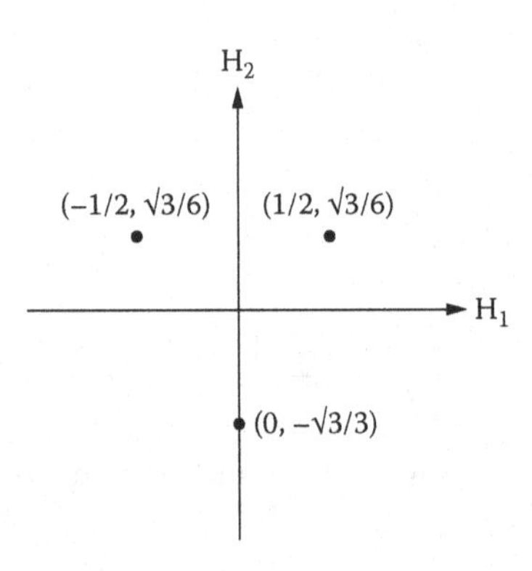

FIGURE 6.1
The conjugate triplet multiplet.

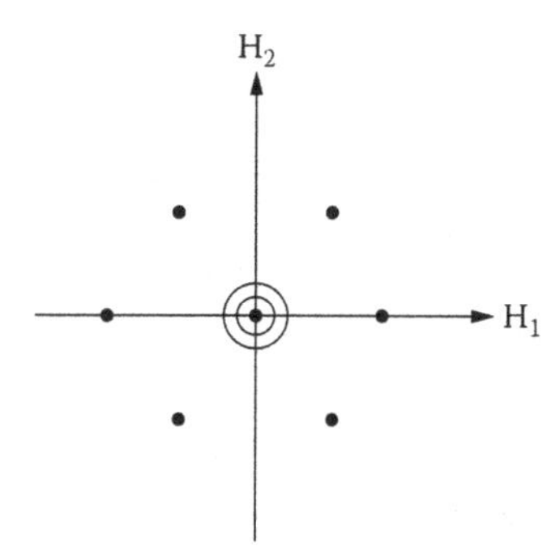

FIGURE 6.2
The octet multiplet and the singlet multiplet.

entry each and can be written as

$$\begin{aligned}
\frac{1}{\sqrt{2}}(T_1 \pm iT_2) &= E_{\pm 1,0} \\
\frac{1}{\sqrt{2}}(T_4 \pm iT_5) &= E_{\pm\frac{1}{2},\pm\frac{\sqrt{3}}{2}} \\
\frac{1}{\sqrt{2}}(T_6 \pm iT_7) &= E_{\pm\frac{1}{2},\mp\frac{\sqrt{3}}{2}}
\end{aligned} \tag{6.20}$$

where the plus and minus signs are correlated. The roots form a regular hexagon, plotted along with the two elements of the Cartan subalgebra in the center (Figure 6.2).

Simple Roots

To complete the analogy between $SU(2)$ and an arbitrary simple Lie algebra we need an idea of positivity for the weights. We can then treat raising and lowering operators and the highest weight. If every nonzero weight is either positive or negative we also know that if μ is positive then $(-)\mu$ is negative.

In an arbitrary basis for the Cartan subalgebra the components (μ, etc.) of the weight arc are fixed. We decide that the weight is positive if its first nonzero component is positive and vice versa. (It actually does not matter what the basis is, but it feels better.)

In $SU(3)$, the three-dimensional defining representation then has a negative weight in the upper left-hand part, a positive weight in the upper right-hand part, and again, has a negative weight in the lower half. We define an ordering by $\mu > \nu$ if $(\mu - \nu)$ is positive and can now think of the highest weight in a representation. You may enjoy working out the $SU(3)$ adjoint weights yourself.

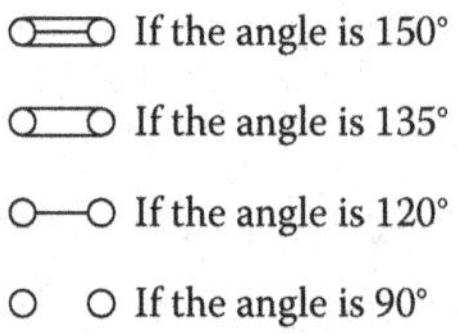

O≡O If the angle is 150°

O=O If the angle is 135°

O—O If the angle is 120°

O O If the angle is 90°

FIGURE 6.3
The four angles between the pairs of simple roots.

We define *simple roots* as positive roots that cannot be made out of others. If a weight is annihilated by all the generators of the simpler roots, then it is the highest weight of an irreducible representation. Now, from the geometry of the simple roots you can reconstruct the whole algebra. I advise students to do this on their own. Using Equation 6.13 and noting that if α and β are different simple roots, then $\alpha - \beta$ is not, and thus, $|E_\alpha >$ is annihilated by $E_{-\beta}$ and $|E_\beta >$ is annihilated by $E_{-\alpha}$; so you can show that

$$\frac{\alpha \cdot \beta}{\alpha^2} = (-)\frac{p}{2} \tag{6.21}$$

$$\frac{\beta \cdot \alpha}{\beta^2} = (-)\frac{p'}{2}. \tag{6.22}$$

If you know the integers p and p' for each simple root then you know the angles between the simple roots and their relative lengths. Indeed

$$\cos(\theta_{\alpha\beta}) = \frac{(-)\sqrt{pp'}}{2} \tag{6.23}$$

and the angle between any pair of simple roots satisfies

$$\frac{\pi}{2} \leq \theta \leq \pi \tag{6.24}$$

where the first inequality follows from Equation 6.23 because the cosine is less than or equal to zero and the second inequality then follows because all the roots are positive.

A *Dynkin diagram* is a shorthand notation for writing the simple roots. Each simple root is shown as an open circle. Pairs of circles are connected by lines, depending on the angle between the pair of roots to which the circles correspond. The scheme is shown in Figure 6.3.

The Dynkin diagram determines all the angles between pairs of simple roots. There may, however, be choices for the relative lengths. We note that Figure 6.4 is the diagram for $SU(2)$ and that Figure 6.5 is the diagram for $SU(3)$.

O is the diagram for $SU(2)$

FIGURE 6.4
The $SU(2)$ multiplet.

O—O Is the diagram for SU(3)

FIGURE 6.5
The $SU(3)$ multiplet.

The Cartan Matrix

You do not need to use dreadful geometrical calculations to keep track of the integers p^i and q^i associated with the action of a simple root α^i on a state $|\phi>$ for a positive root ϕ. The idea is to label the roots directly by their $q^i - p^i$ values. The $q^i - p^i$ of any weight, μ, is simply twice its E_3 value, where E_3 is the Cartan generator of the $SU(2)$ associated with the simple root α^i. Because

$$2E_3|\mu> = \frac{2H \cdot \alpha^i}{(\alpha^i)^2}|\mu> = (q^i - p^i)|\mu>, \tag{6.25}$$

if A is the *Cartan matrix* then using Equation (6.25) and the form A of the Cartan matrix

$$A_{ji} = \frac{2\alpha^j \cdot \alpha^i}{(\alpha^i)^2} \tag{6.26}$$

will give us all the same information as the Dynkin diagram. For $SU(3)$, the Cartan matrix has the form

$$\begin{pmatrix} 2 & -1 \\ -1 & 2 \end{pmatrix}. \tag{6.27}$$

Finding All the Roots

We can use the Cartan matrix to simplify calculating all of the roots from the simple roots.

The action of the raising operator $\mathbf{E}_{\alpha j}$ moves ϕ to $\phi + \alpha^j$. This just changes k_j to k_{j+1} and thus $q^i - p^i$ to $q^i - p^i + A^{ji}$, that is,

$$\begin{aligned} k_i &\to k_j + 1, \\ q^i - p^i &\to q^i - p^i + A^{ji}\,. \end{aligned} \tag{6.28}$$

If we think of the $q^i - p^i$ as the elements of a row vector, this is equivalent to simply adding the jth row of the Cartan matrix, which is simply the vector $q - p$ associated with the simple root α^i. This speeds up the calculations of the roots. We will do it for $SU(3)$.

Start with the simple roots in the $q - p$ notation. Put each in a rectangular box and arrange them on a hortizontal line representing the $k = 1$ layer of positive roots, that is, the simple roots.

$$k = 1 \quad \boxed{2\text{-}1} \quad \boxed{\text{-}1\ 2} \quad \alpha^1, \alpha^2. \tag{6.29}$$

Now put a box with m zeros, representing the Cartan generators on a line below, representing the $k = 0$ layer.

$$k = 0 \quad \boxed{0\ 0} \quad H_i. \tag{6.30}$$

Now for each element of each box we know the q^i value. For the ith element of α^i, $q^i = 2$ because the root is part of the $SU(2)$ spin 1 representation, consisting of $E_{\pm}\alpha^i$ and $\alpha^i \cdot H$. For all the other elements, $q^i = 0$ because $\alpha^i - \alpha^j$ is not a root.

Thus

$$\begin{array}{lccc} & q = 2\ \ 0 & 0\ \ 2 & \\ k = 1 & \boxed{2\ \text{-}1} & \boxed{\text{-}1\ 2} & \alpha^1, \alpha^2 \\ k = 0 & \boxed{0\ 0} & & H_i. \end{array} \tag{6.31}$$

We can compute the corresponding p^i.

$$\begin{array}{lccc} & p = 0\ \ 1 & 1\ \ 0 & \\ k = 1 & \boxed{2\ \text{-}1} & \boxed{\text{-}1\ 2} & \alpha^1, \alpha^2 \\ k = 0 & \boxed{0\ 0} & & H_i. \end{array} \tag{6.32}$$

Since the ith element of α^i is 2, the corresponding p^i is zero. For all the others, p is just minus the entry. For each nonzero p, we draw a line from the simple root to a new root with $k = 2$ on a horizontal line above the $k = 1$ line, obtained by adding the appropriate simple root. You can also draw such lines from the $k = 0$ layer to the $k = 1$ layers and the lines for each root will have a different angle. Then try to put the boxes on the $k = 2$ layer so that the lines associated with each root have the same angle they did between the 0 and 1 layer. These lines represent the action of the $SU(2)$ raising and lowering operators.

$$p = 0\ 0$$
$$q = 0\ 0$$

$$\begin{array}{lccl} k=2 & & \boxed{1\,1} & \alpha^1, \alpha^2 \\ k=1 & \boxed{2\,\text{-}1} & \boxed{\text{-}1\,2} & \alpha^1, \alpha^2 \\ k=0 & & \boxed{0\,0}. & \end{array} \tag{6.33}$$

This is now trivial to iterate, for everything you need to go from $k = 1$ to $k = 1 + 1$ is on the diagram. For $SU(3)$, the procedure terminates at $k = 2$ because all the p's are zero.

$$\begin{array}{lccl} k=2 & & \boxed{1\,1} & \alpha^1 + \alpha^2 \\ & & / \quad \backslash & \\ k=1 & \boxed{2\,1} & \boxed{\text{-}1\,2} & \alpha^1, \alpha^2 \\ & & \backslash \quad / & \\ k=0 & & \boxed{0\,0} & H_i \\ & & / \quad \backslash & \\ k=-1 & \boxed{1\,\text{-}2} & \boxed{\text{-}2\,1} & -\alpha^2, -\alpha^1 \\ & & \backslash \quad / & \\ k=-2 & & \boxed{\text{-}1\,\text{-}1} & -\alpha^1 - \alpha^2 \end{array} \tag{6.34}$$

Fundamental Weights

Suppose that the simple roots of some Lie algebra are α^j from $j = 1$ to m. The highest weight, μ, of an arbitrary irreducible representation, D, has the property that $\mu + \phi$ is not a weight for any positive root ϕ. Thus, $\mu + \alpha^j$ not to be a weight in the representation j is clearly sufficient, because then

$$E_j^{\alpha}|\mu> = 0 \text{ for all } j, \tag{6.35}$$

which implies that all positive roots annihilate the state. This is clearly an if-and-only-if statement. Thus, μ is the highest weight of an irreducible representation. Hence, for every E_j^{α} acting on $|\mu> p = 0$ and thus

$$\frac{2\alpha^j \cdot \mu}{(\alpha^j)^2} = \ell^j \tag{6.36}$$

where the ℓ^j are nonnegative integers. The ℓ^j completely determine μ. Every set of ℓ^j gives μ, which is the highest weight of some irreducible representations. Hence the irreducible representations of rank m simple Lie algebras can be labeled by a set of m nonnegative integers ℓ^j. These integers are called the *Dynkin coefficients.*

Consider the weight vectors, μ^j, satisfying

$$\frac{2\alpha^j \cdot \mu^k}{(\alpha^j)^2} = \delta_{jk}. \tag{6.37}$$

Every highest weight can be written uniquely as

$$\mu = \sum_{j=1}^{m} \ell^j \mu^j. \tag{6.38}$$

The vectors μ^j are called the *fundamental weights* and the m irreducible representations that have these as highest weights are called the *fundamental representations.* We often denote them by D^j. Note that the superscripts are just labels. The vectors also have vector indices. (Both run from 1 to m—confusing.)

Now running the previous arguments backward gives

$$\ell^j = q^j - p^j, \tag{6.39}$$

that is, ℓ^j is the $q^j - p^j$ value of the simple root α^j.

The Weyl Group

There is a symmetry that appears because there is an $SU(2)$ associated with each root direction and all $SU(2)$ representations are symmetrical under the reflection $E_3 \to (-)E_3$. If μ is a weight and $E_3 \equiv \alpha \cdot \frac{H}{\alpha^2}$ is the E_3 associated with the root α, then

$$E_3|\mu> = \frac{\alpha \cdot \mu}{\alpha^2}|\mu> \tag{6.40}$$

and the reflection symmetry implies that $\mu - (q - p)\alpha$ (where $q - p = 2(\frac{\alpha \cdot \mu}{\alpha^2})$) is a weight with the opposite E_3 value. There are reflections for all roots that are transformations on the weight space and that leaves the roots unchanged.

The set of all such transformations obtained in this way forms a transformation group called the *Weyl group* of the algebra. The individual reflections are called Weyl reflections. The Weyl group is a simple way of understanding the hexagonal and triangular structures that appear in $SU(3)$ representations.

Young Tableaux

You may have met Young tableaux in discussions of irreducible representations of the symmetric groups. We will now see that they are useful for dealing with the irreducible representations of Lie groups. We will begin by discussing this for $SU(3)$ but the real advantage is that it generalizes to higher groups.

Raising the Indices

The crucial observation is that the $\overline{3}$ representation is an antisymmetric combination of two 3 representations, so we do not need the second fundamental representation to construct higher representations. We can write an arbitrary representation as a tensor product of 3's with appropriate symmetry. In fact, irreducible representations of $SU(3)$ transform irreducibly under permutation of the indices. Consider a general representation (n, m). It is a tensor with components

$$A = \begin{matrix} i_1 \ldots i_n \\ \\ j_1 \ldots j_m \end{matrix}$$

separately symmetric in upper and lower indices and traceless. We can raise all the lower indices with ε tensors to get

$$a^{i_1 \ldots i_n k_1 l \ldots k_m l_n} = \varepsilon^{j_1 k_2 l_3 \ldots} \varepsilon^{j_m k} A^{\ell_i \ldots \ell_n}_{j_i \ldots j_m}. \tag{6.41}$$

Clearly, it is antisymmetric in each pair, $k_i \leftrightarrow \ell_i$, and antisymmetric in the exchange of pairs, $k_i, \ell_c \leftrightarrow k_j \ell_j$. Now for each such tensor, we can associate a Young tableau (Figure 6.6).

Think about the highest weight of the representation, (n, m). Because the lowering operators preserve the symmetry, if we find the symmetry of the tensor components describing the highest weight, all the states will have that symmetry. The highest weight is associated with the components in which all there is one 1 and all the k, ℓ pairs are 1, 3. All of these can be obtained by antisymmetrizing the k, ℓ pairs from the component in which all the k's

k_1	...	k_m	i_i	...	i_n
l_1	...	l_m			

FIGURE 6.6
The Young tableau of the (m, n) representation.

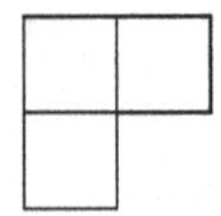

FIGURE 6.7
The empty Young tableau of the (2, 1) representation.

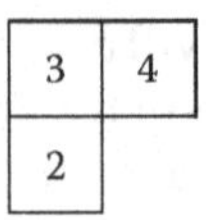

FIGURE 6.8
The eight-dimensional Young tableau.

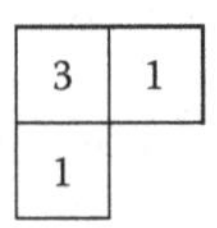

FIGURE 6.9
The three-dimensional Young tableau.

are one, and all the ℓ's are three. But this one component is symmetric under arbitrary permutations of the ℓ's. Thus, we will obtain a tensor with the right symmetry if we start with an arbitrary tensor with $n+2m$ components and first symmetrize all the i's and k's and separately the ℓ's, and then antisymmetrize in every k, ℓ pair.

In the Young tableau language we first symmetrize in the components in the rows then antisymmetrize in the components in the columns. The result is symmetric in the i's and in the k, ℓ pairs. It is also traceless.

In $SU(3)$ the tensors corresponding to Young tableaux with more than three boxes in any column vanish. There are simple rules that allow us to calculate the dimension of multiplets from the corresponding Young tableaux. Consider the tableau in $SU(3)$. First we start with the 3 of $SU(3)$ in the top left corner and increase across the rows and decrease down the columns by unity in both cases (Figure 6.8). This works for any $SU(n)$. Then multiply these numbers, so that we get 24.

Now put the hook lengths into the boxes where the hook is up and to the right with the corner in the relevant box. Here we get the result in Figure 6.9 and multiplying gives 3. The dimension of the multiplet is the quotient of these two numbers. Here, 24 divided by 3 gives 8.

Similarly, methods work for all the Lie groups except for the exceptional ones that do not coincide with nonexceptional ones.

The Classification Theorem (Dynkin)

See chapter 20, p. 244 of *Lie Algebras in Particle Physics* (2nd ed.) by H. Georgi.

Result

Coincidences

1. A_1, B_1, C_1 are all $SU(2)$.
2. $B_2 = C_2$.
3. $D_3 = A_3$.
4. Remove one more circle from D_3 to get D_2 and it falls apart into two disconnected circles (the middle one must be removed to stay in the D_n family). Thus, D_2 is not simple. This is the important statement that the algebra of $SU(4)$ is the same as the algebra of $SU(2) \times SU(2)$.

This is the complete list of such coincidences.

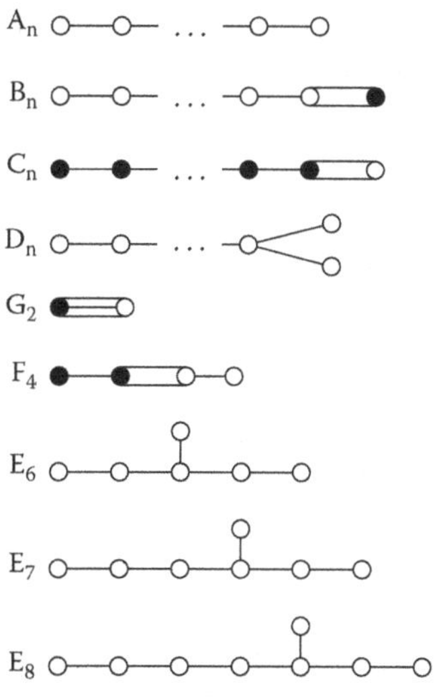

FIGURE 6.10
The four general dimensional Dynkin classes and the five exceptional Dynkin classes.

References

1. H. Georgi, *Lie Algebras in Particle Physics*, 2nd ed., Westview Press, Boulder, CO, 1999.
2. J.F. Cornwell, *Group Theory in Physics*, Vols. I and II, Academic Press, San Diego, CA, 1984.
3. N.E.P. Samara and R.C. King, *J. Phys. A. Math. Gen.* 12, no. 12 (1979): 2315.
4. N.E.P. Samara and R.C. King, *J. Phys. A. Math. Gen.* 12, no. 12 (1979): 2317.

Problems

6.1 Show that $[E_\alpha, E_{-\alpha} = \alpha.H]$ which should remind you of the $SU(2)$ commutation relation $[J^+, J^-] = J_3$. This analogy will be exploited to learn more about the representations of compact Lie groups. Generally, for any weight μ of a representation D, the E_3 value is given by $E_3|\mu, x, D >= \frac{\alpha\cdot\mu}{\alpha^2}|\mu, x, D >$. Because the E_3 values must be integers or half integers, $\frac{2\alpha\cdot\mu}{\alpha^2} =$ integer.

6.2 Show that

$$\frac{\alpha \cdot \mu}{\alpha^2} = -\frac{1}{2}(p - q)$$

where $\mu + p\alpha$ is the weight of the highest E_3 state of the $SU(2)$ spin j representation and q plays the corresponding role when lowering.

6.3 We can easily make use of these results. Apply Equation (6.13) to two distinct roots α and β and we find (using E_α as the $SU(2)$ definition)

$$\frac{\alpha \cdot \beta}{\alpha^2} = (-)\frac{1}{2}(p - q).$$

Again, now using E_β yields instead

$$\frac{\beta \cdot \alpha}{\beta^2} = (-)\frac{1}{2}(p' - q').$$

6.4 If $\theta_{\alpha\beta}$ is the angle between the roots α and β then show that multiplying the last two results gives

$$\cos^2(\theta_{\alpha\beta}) = \frac{(\alpha \cdot \beta)^2}{\alpha^2\beta^2} = \frac{(p - q)(p' - q')}{4}.$$

6.5 These vectors, plotted in a plane, form the vertices of an equilateral triangle. Show that this is true.

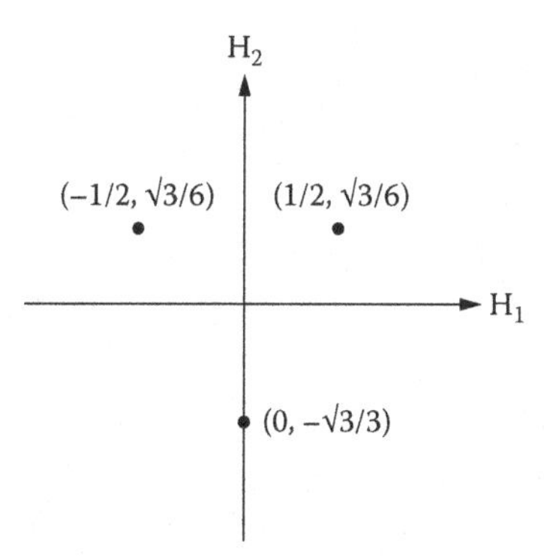

6.6 The roots are differences of weights, because the corresponding generators take one weight to another. The generators clearly have only one-off diagonal entry each and can be written as

$$\frac{1}{\sqrt{2}}(T_1 \pm i T_2) = E_{\pm 1,0}$$
$$\frac{1}{\sqrt{2}}(T_4 \pm i T_5) = E_{\pm\frac{1}{2},\pm\frac{\sqrt{3}}{2}}$$
$$\frac{1}{\sqrt{2}}(T_6 \pm i T_7) = E_{\pm\frac{1}{2},\mp\frac{\sqrt{3}}{2}}$$

where the plus and minus signs are correlated. The roots form a regular hexagon, plotted along with the two elements of the Cartan subalgebra in the center:

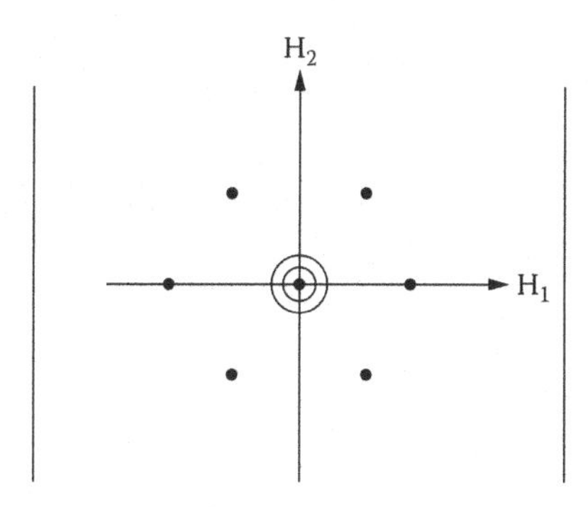

6.7 Show that

$$\frac{\alpha \cdot \beta}{\alpha^2} = (-)\frac{p}{2}.$$

6.8 Show that

$$\frac{\beta \cdot \alpha}{\beta^2} = (-)\frac{p'}{2}.$$

6.9 The reader may enjoy the simple exercise of showing that the simple roots are linearly independent. By exploiting the completeness of the simple roots, it is easy to find all of them.

6.10

$$2E_3|\mu> = \frac{2H \cdot \alpha^i}{(\alpha^i)^2}|\mu> = (q^i - p^i)|\mu>,$$

if A is the Cartan matrix then using Equation (6.30) and the form A of the Cartan matrix

$$A_{ji} = \frac{2\alpha^j \cdot \alpha^i}{(\alpha^i)^2}$$

will give us all the same information as the Dynkin diagram. For $SU(3)$, the Cartan matrix has the form

$$\begin{pmatrix} 2 & -1 \\ -1 & 2 \end{pmatrix}.$$

6.11 The action of the raising operator $\mathbf{E}_{\alpha j}$ moves ϕ to $\phi + \alpha^j$. This just changes k_j to k_{j+1} and thus $q^i - p^i$ to $q^i - p^i + A^{ji}$, that is $k_i \to k_j + 1$.

6.12 Start with the simple roots in the $q - p$ notation. Put each in a rectangular box and arrange them on a hortizontal line representing the $k = 1$ layer of positive roots, that is, the simple roots.

$k = 1$ | 2 -1 | | -1 2 | α^1, α^2.

Now put a box with m zeros, representing the Cartan generators, on a line below, representing the $k = 0$ layer.

$k = 0$ | 0 0 | H_i.

Now for each element of each box we know the q^i value. For the ith element of α^i, $q^i = 2$ because the root is part of the $SU(2)$ spin 1 representation, consisting of $E_{\pm}\alpha^i$ and $\alpha^i \cdot H$. For all the other elements, $q^i = 0$ because $\alpha^i - \alpha^j$ is not a root.

6.13 Thus

$q = 2\ 0 \quad 0\ 2$

$k = 1$ | 2 -1 | | -1 2 | α^1, α^2

$k = 0$ | 0 0 | H_i.

6.14 We can compute the corresponding p^i.

$p = 0\ 1 \quad 1\ 0$

$k = 1$ | 2 -1 | | -1 2 | α^1, α^2

$k = 0$ | 0 0 | H_i.

6.15 Hence, for every E_j^α acting on $|\mu>$ $p = 0$ and thus

$$\frac{2\alpha^j \cdot \mu}{(\alpha^j)^2} = \ell^j$$

where the ℓ^j are nonnegative integers. The ℓ^j completely determine μ. Every set of ℓ^j gives μ, which is the highest weight of some irreducible representations. Hence the irreducible representations of rank m simple Lie algebras can be labeled by a set of m nonnegative integers ℓ^j. These integers are called the Dynkin coefficients. Consider the weight vectors, μ^j, satisfying

$$\frac{2\alpha^j \cdot \mu^k}{(\alpha^j)^2} = \delta_{jk}.$$

Every highest weight can be written uniquely as

$$\mu = \sum_{j=1}^{m} \ell^j \mu^j.$$

6.16 Now running the previous arguments backward gives

$$\ell^j = q^j - p^j$$

that is, ℓ^j is the $q^j - p^j$ value of the simple root α^j.

6.17 The crucial observation is that the $\overline{3}$ representation is an antisymmetric combination of two 3 representations, so we do not need the second fundamental representation to construct higher representations. Show in your own notation how to do this.

6.18 Now for each such tensor, we can associate a Young tableau. In your own words show how this works.

6.19 In $SU(3)$ the tensors corresponding to Young tableaux with more than three boxes in any column vanish. There are simple rules that allow us to calculate the dimension of multiplets from the corresponding Young tableaux. Consider the tableau

<table>
<tr><td>k_1</td><td>...</td><td>k_m</td><td>i_i</td><td>...</td><td>i_n</td></tr>
<tr><td>l_1</td><td>...</td><td>l_m</td></tr>
</table>

in $SU(3)$. In your own notation show how this works.

7

Noether's Theorem and Gauge Theories of the First and Second Kinds

Perhaps the main point of Lagrangian formalism is that it provides a natural framework for the quantum mechanical implementation of symmetries. This is caused by the principle of stationary action taking the form of a variational principle in the dynamical equations of the Lagrangian formalism. Consider any infinitesimal transformation of the fields

$$\Psi^k(\underline{x}) \rightarrow \Psi^k(\underline{x}) + i\varepsilon F^k(\underline{x}) \tag{7.1}$$

which leaves the action

$$I[\Psi] \equiv \int_{-\infty}^{\infty} dt L[\Psi(t), \dot{\Psi}(t)] \tag{7.2}$$

invariant. Under an arbitrary variation of $\Psi(x)$ we get

$$\delta I[\Psi] = \int_{-\infty}^{\infty} dt \int d^3\underline{x} \left[\frac{\delta L}{\delta \Psi^k} \delta\Psi^k(\underline{x}) + \frac{\delta L}{\delta \dot{\Psi}^k} \delta\dot{\Psi}^k(x) \right]. \tag{7.3}$$

Now assume that $\delta\Psi^k(\underline{x})$ vanishes for $t \rightarrow \pm\infty$ so that we may integrate by parts, and write

$$\delta I[\Psi] = \int d^4x \left[\frac{\delta L}{\delta \Psi^k(\underline{x})} - \frac{d}{dt} \frac{\delta L}{\delta \dot{\Psi}(\underline{x})} \right] \delta\Psi^k(\underline{x}) . \tag{7.4}$$

We see that the action is stationary with respect to all variations $\delta\Psi^k$ that vanish at $t \rightarrow \pm\infty$ if and only if the field satisfies the field equations

$$\dot{\Pi}_k(\underline{x}, t) = \frac{\partial L[\Psi(t), \dot{\Psi}(t)]}{\delta \Psi^k(\underline{x}, t)} . \tag{7.5}$$

Notice that we could have come at this from a different, and perhaps more familiar, point starting with a Lagrangian as a function of a set of generic fields $\Psi^k(\underline{x}, t)$ and their time derivatives $\dot{\Psi}(\underline{x}, t)$, when the conjugate fields $\Pi_k(\underline{x}, t)$ are defined as the variational derivatives.

DOI: 10.1201/9781439895207-7

For all such schemes we can introduce the Lagrangian density, $\mathcal{L}$, a scalar function of $\Psi(x)$ and

$$\frac{\partial\Psi(x)}{\partial x^{\mu}}$$

so that the action is

$$I[\Psi] = \int d^4x \ \mathcal{L}\left(\Psi(x), \frac{\partial\Psi(x)}{\partial x^{\mu}}\right). \tag{7.6}$$

All field theories used in current theories of elementary particles have Lagrangians of this form. Varying $\Psi^k(x)$ by an amount $\delta\Psi^k(x)$ and integrating by parts we find the variation in L is

$$\delta L = \int d^3x \left[\left(\frac{\partial\mathcal{L}}{\partial\Psi^k} - \underline{\nabla}\cdot\frac{\partial\mathcal{L}}{\partial\underline{\nabla}\Psi^k}\right]\delta\Psi^k + \frac{\partial\mathcal{L}}{\partial\Psi^k}\partial\dot{\Psi}^k\right], \tag{7.7}$$

so that with obvious arguments suppressed we get

$$\frac{\delta L}{\partial\Psi^k} = \frac{\partial\mathcal{L}}{\partial\dot{\Psi}^k}. \tag{7.8}$$

The field equations now read

$$\frac{\partial}{\partial x^{\mu}}\frac{\partial\mathcal{L}}{\partial(\partial\Psi^k)(\partial x^{\mu})} = \frac{\partial\mathcal{L}}{\partial x^k}. \tag{7.9}$$

These are known as the Euler–Lagrange equations. As expected, if $\mathcal{L}$ is a scalar, these equations are Lorentz invariant. In addition, t being Lorentz invariant, the action I is required to be real. This is because we want just as many field equations as there are fields. This reality condition also ensures that the generators of various symmetry transformations are Hermitian operators.

We now come to the real point of the Lagrangian formalism—that it provides a natural framework for the quantum mechanical interpretation of symmetry principles. This is because the dynamical equations in the Lagrangian formalism take the form of a variational principle, the principle of stationary action. Consider any infinitesimal transformation of the fields

$$\Psi^k(x) \rightarrow \Psi^k(x) + i\varepsilon F^k(x) \tag{7.10}$$

that leaves the action invariant. If ε is constant, such symmetries are known as *global symmetries*. Of course, the action is invariant if the fields satisfy the dynamical equations. By an infinitesimal symmetry transformation we mean one that leaves the action invariant even when the dynamical equations are not satisfied. Now consider the same transformation with ε an arbitrary function of position in space–time, then, in general, the variation of the action will not vanish. But it will have to be of the form

$$\delta I = (-)\int d^4x J^{\mu}(x)\frac{\partial\varepsilon(x)}{\partial x^{\mu}} \tag{7.11}$$

in order that it should vanish when $\varepsilon(x)$ is constant. If we now take the fields in $I(\Psi)$ to satisfy the field equations, then I is stationary with respect to arbitrary field variations that vanish at large space–time distances. These include variations of the form in Equation (7.10), so in this case δI should vanish. Integrating by parts, we see that $J^{\mu}(x)$ must satisfy a conservation law

$$\frac{\partial J^{\mu}(x)}{\partial x^{\mu}} = 0 . \tag{7.12}$$

It follows that $\frac{dF}{dt} = 0$

$$\text{where} \quad F \equiv \int d^3x J^0 . \tag{7.13}$$

There is one such conserved current J^{μ} and one constant of the motion F for each independent infinitesimal symmetry transformation. This represents a general feature of the canonical formalism, often referred to as Noether's theorem: *symmetries imply conservation laws.* This theorem [1] is cited in the original German and in the English translation, which Einstein is known to have encouraged strongly. (Note that this theorem is by a woman author working alone when such things were far from easy.)

Now we turn our attention to the treatment of first and second class restraints and Dirac brackets [2]. The main problem to deriving the Hamiltonian from the Lagrangian is the presence of constraints. *Primary constraints* are either imposed on the system (a good example is in picking a gauge for the electromagnetic field) or arise from the structure of the Lagrangian itself. A good example is found by considering the Lagrangian of a massive vector field V^{μ} interacting with a current J_{μ} where we have

$$\mathcal{L} = (-)\frac{1}{4}F_{\mu\nu}F^{\mu\nu} - \frac{1}{2}m^2 V_{\mu}V^{\mu} + J_{\mu}V^{\mu} \tag{7.14}$$

$$\text{where} \quad F_{\mu\nu} \equiv \partial\mu V_{\nu} - \partial_{\nu} V_{\mu} . \tag{7.15}$$

To treat all indices on the same basis we define the conjugates

$$\Pi^{\mu} \equiv \frac{\partial\mathcal{L}}{\partial(\partial_0 V_{\mu})} = (-)F^{0\mu} . \tag{7.16}$$

We find the primary constraint

$$\Pi_0 = 0 . \tag{7.17}$$

Primary constraints are found when the equations

$$\Pi_{\ell} = \frac{\delta L}{\delta\partial_0\psi^{\ell}} \tag{7.18}$$

cannot be solved to give all the $\partial_0 \Psi^\ell$ (at least locally) in terms of Π_ℓ and Ψ^ℓ. This will be the case if and only if the matrix of the two first partial time derivatives of the Lagrangian has a vanishing determinant. Such Lagrangians are called irregular.

Then there are *secondary constraints*, which arise from the requirement that the primary constraints be consistent with the equations of motion. For the massive vector field, this is just the Euler–Lagrange equation for V_0

$$\partial_i \Pi_i = m^2 V^0 - J^0 . \tag{7.19}$$

There are many variations on this theme but we do not need them here. Much more important for us is the distinction between first and second type restraints. The constraints we have found for the massive vector field are of a type known as second class, for which there is a universal prescription for commutation relations.

To explain the distinction between first and second class restraints, we recall the definition of the Poisson brackets of classical mechanics. Consider any Lagrangian $L(\Psi, \dot{\Psi})$ that depends on a set of variables $\Psi^a(t)$ and their time derivatives $\dot{\Psi}^a(t)$. We define canonical conjugates for all of these variables by

$$\Pi_a \equiv \frac{\partial L}{\partial \dot{\Psi}^a} . \tag{7.20}$$

The Πs and Ψs will in general not be independent variables, but may instead be related by various constraint equations, both primary and secondary. The Poissson bracket is then defined by

$$[A, B] \equiv \frac{\partial A}{\partial \Psi^a} \frac{\partial B}{\partial \Pi_a} - \frac{\partial B}{\partial \Psi^a} \frac{\partial A}{\partial \Pi_a} , \tag{7.21}$$

with the constraints ignored in calculating the derivatives. In particular, we always have

$$[\Psi^a, \Pi_b] = \delta^a_b \tag{7.22}$$

where from now on all fields are taken at the same time and time arguments are everywhere dropped. We call a constraint *first class* if its Poisson bracket with all the other constraints vanishes when (after calculating the Poisson brackets) we impose the constraints. Such constraints can always be eliminated by a choice of gauge.

After all of the first class constraints have been eliminated by a choice of gauge, the remaining constraint equations

$$X_n = 0 \tag{7.23}$$

are such that no linear combination of the Poisson brackets of these constraints with each other vanishes. It follows that the matrix C of the Poisson brackets of the remaining constraints is nonsingular:

$$DetC \neq 0 \tag{7.24}$$

where

$$C_{NM} \equiv [X_N, X_M]_P \;. \tag{7.25}$$

Constraints of this sort are called *second class*. There must always be an even number of second class restraints, because an antisymmetric matrix of odd dimensionality has to have a vanishing determinant.

Dirac suggested that when all constraints are second class, the commutators will be given by a simple modification when he called the resulting Poisson bracket the **Dirac bracket**. A powerful theorem by Maskuwa and Nakajima [3] was used to examine the Dirac bracket and its properties, but the issue appears to remain unresolved.

References

1. E.M. Noether, *Transport Theory Statist. Phys.* 1 no. 3, (1971): 183; E.M. Noether, Invariant variation problems. *Nachr. Acad. Wess., Götingen, Math-Phys. Kl.* II (1918): 235.
2. P.A.M. Dirac, Lectures on Quantum Mechanics, Yeshiva University, New York, 1964. Also see P.A.M. Dirac, *Can. J. Math.* 2 (1950): 1929; *Proc Roy. Soc. London*, ser A, 246 (1958): 326.
3. T. Maskawa and H. Nakajima, *Prog. Theor. Phys.* 56 (1976): 1295.

Problems

7.1 Read from Equation (7.1) through Equation (7.5). Close the book. Now write out your own version of this section.

7.2 Read the section immediately after Equation (7.5). Follow the suggestion and work out the calculation in this alternative manner.

7.3 Using the action in Equation (7.6), integrate by parts to find the Euler–Lagrange equations as given in Equation (7.9).

7.4 In your own words explain what you understand by the principle of stationary action.

7.5 Starting from Equation (7.10) work, with the book closed, until you reach the result in Equation (7.14).

7.6 Express the result of Problem 7.5 in simple English.

7.7 Starting from the definition of the Poisson brackets of classical mechanics, explain what you understand by primary and secondary constraints.

7.8 Explain why there must always be an even number of second class restraints.

7.9 Explain what Dirac meant by the Dirac bracket.

7.10 Look up the paper by Maskawa and Nakajima. Explain in your own terms anything you understand.

8

Basic Couplings of the Electromagnetic, Weak, and Strong Interactions

We start yet again with electromagnetism, which we examined in some detail in Chapter 7. Since A^0 is not an independent Heisenberg-picture field variable we do not introduce any corresponding operator a^0 in the interaction picture, but rather take

$$a^0 = 0\,. \tag{8.1}$$

The most general real solution may be written

$$\begin{aligned} a^\mu(x) = (2\pi)^{\frac{-3}{2}} \int \frac{d^3p}{\sqrt{2p^0}} \sum_\sigma \big[& e^{ip.x} e^\mu(\mathbf{p},\sigma) a(\mathbf{p},\sigma) \\ & + e^{-ip.x} e^\mu(\mathbf{p},\sigma) a^\dagger(\mathbf{p},\sigma) \big]\,. \end{aligned} \tag{8.2}$$

We can easily see that if (and in fact only if) the operator coefficients in Equation (8.2) satisfy

$$[a(p,\sigma)], \sigma^\dagger(p',\sigma')] = \delta^3(p-p')\delta\sigma\sigma' \tag{8.3}$$

$$[a(p,\sigma), a(p',\sigma')] = 0\,. \tag{8.4}$$

The free photon Hamiltonian takes the expected form. The general Feynman rules yield

$$(-i)\Delta_{\mu\nu}(x-y) = \int \frac{d^3p}{(2\pi^3)2|\mathbf{p}|} P_{\mu\nu}(\mathbf{p})[e^{ip.(x-y)}\theta(x-y) + e^{ip(y-x)}\theta(y-x)], \tag{8.5}$$

where

$$P_{\mu\nu}(\mathbf{p}) \equiv \sum_{\sigma=\pm 1} e_\mu(p\sigma) e_\nu(p,\sigma) \tag{8.6}$$

and p^μ in experiments is taken with

$$p^0 = |\mathbf{p}|\,. \tag{8.7}$$

DOI: 10.1201/9781439895207-8

From a practical point of view, the important thing is that in the momentum space Feynman rules, the contribution of an internal photon line is simply given by

$$\frac{(-i)\eta_{\mu\nu}}{(2\pi)^4 q^2 - i\varepsilon} \tag{8.8}$$

and the Coulomb interaction is dropped. (We have not even given a hand-waving argument for this.) It can be justified by a detailed analysis of Feynman diagrams but the easiest way to treat this problem is by path integral methods.

We can now state the Feynman rules for calculating the S-matrix in a quantum electrodynamics. For simplicity we take a single type of spin $\frac{1}{2}$ particles of charge $q = -e$ and mass m. The simplest gauge invariant and Lorentz invariant Lagrangian for this theory is

$$\mathcal{L} = (-)\frac{1}{4}F_{\mu\nu}F^{\mu\nu} - \bar{\Psi}(\gamma^{\mu}(\partial_{\mu} + ieA_{\mu}) + m)\Psi \,. \tag{8.9}$$

The electric current four-vector is then

$$J^{\mu} = \frac{\partial\mathcal{L}}{\partial A_{\mu}} = (-)ie\bar{\Psi}\gamma^{\mu}\Psi \,. \tag{8.10}$$

It should now be obvious to the reader how the spin $\frac{1}{2}$ particles enter using the familiar Pauli matrices.

The reader will also know how in the early 1960s Gell-Mann [1] extended the approximate $SU(2)$ isospin symmetry of nuclear physics to an even less exact $SU(3)$ symmetry, which grouped the partially known baryonic and mesons into octets and decouplets. These are now known as:

- $\frac{1}{2}^{+}$ baryons $p, n, \Omega^0, \Sigma^{+0-}, \Xi^{-1,0}$
- An octet of O^- mesons $K^{+,0}, \Pi^{+0-}, \eta^0, \bar{\kappa}^{-,0}$
- An octet of 1^- mesons $\kappa^{*+,0}\rho'\omega, \bar{\kappa}^{-,0}$
- A decouplet of $\frac{3}{2}^{+}$ baryons $\Lambda^{++,+,0,-}, \Sigma^{*+,0,-1}, \Xi^{*0,-}, \Omega^-$

After the successes of the chiral $SU(2) \times SU(2)$ symmetry in the mid-1960s, it was natural to suppose that the strong interactions also respect an approximate $SU(3) \times SU(3)$ symmetry, which like $SU(2) \times SU(2)$ is spontaneously broken to its diagonal subgroup, the Gell-Mann $SU(3)$. Quantum chromodynamics revealed it arises because of there being not merely two fairly light quarks—the u and the d—but a third one s has the same charge as the d and is still fairly light. This means that the $SU(3) \times SU(3)$ symmetry consists of independent $SU(3)$ transformation on the left- and right-handed parts of the u, d, s quark fields:

$$\begin{pmatrix} u \\ d \\ s \end{pmatrix} \rightarrow exp\left[i\sum_a \left(\theta_a^V \lambda_a + \theta^A \lambda_a \gamma_5\right)\right]\begin{pmatrix} u \\ d \\ s \end{pmatrix} \tag{8.11}$$

where the λ_a are the complete set of traceless Hamiltonian (3×3) matrices:

$$\lambda_1 = \begin{pmatrix} 0 & 1 & 0 \\ 1 & 0 & 0 \\ 0 & 0 & 0 \end{pmatrix}, \quad \lambda_2 = \begin{pmatrix} 0 & -i & 0 \\ i & 0 & 0 \\ 0 & 0 & 0 \end{pmatrix}, \quad \lambda_3 = \begin{pmatrix} 1 & 0 & 0 \\ 1 & -1 & 0 \\ 0 & 0 & 0 \end{pmatrix},$$

$$\lambda_4 = \begin{pmatrix} 0 & 0 & 1 \\ 0 & 0 & 0 \\ 1 & 0 & 0 \end{pmatrix}, \quad \lambda_5 = \begin{pmatrix} 0 & 0 & -i \\ 0 & 0 & 0 \\ i & 0 & 0 \end{pmatrix}, \quad \lambda_6 = \begin{pmatrix} 0 & 0 & 0 \\ 0 & 0 & 1 \\ 0 & 1 & 0 \end{pmatrix},$$

$$\lambda_7 = \begin{pmatrix} 0 & 0 & 0 \\ 0 & 0 & -i \\ 0 & i & 0 \end{pmatrix}, \quad \lambda_8 = \frac{1}{\sqrt{3}}\begin{pmatrix} 1 & 0 & 0 \\ 0 & 1 & 0 \\ 0 & 0 & -2 \end{pmatrix} \tag{8.12}$$

with $Tr.(\lambda_a \lambda_b) \equiv 2\delta_{ab}$ as normalization. This general scheme does seem in one way or another to have continued to expand, so that we now have also a charmed quark c, a bottom (or beautiful) quark b, and the top quark t, which appears to be the heaviest of them all. The reader is warned that there are other patterns in which quarks could have emerged, with very different consequences. One such scheme is that of Barnes, Jarvis, and Ketley [2] where instead of quarks, and possible partners, appearing as representing multiplets of a given $SU(2)$, new $SU(2)$s appear with new sets of quarks and possible partners. After charm, the next one in this scheme is called style, stealing from a Frank Sinatra song with the snatches "You either have or you haven't got style, if you have it sticks out a mile" and "Style and charm kind of go arm in arm."

Returning to the usual notation, by the mid-1960s, it was understood that the weak interaction processes of hadrons with each other and with leptons are well described at low energy by the effective Lagrangian

$$\frac{G_F}{\sqrt{2}}\,[\bar{e}\gamma(1 + i\gamma_5)\nu_e + \bar{u}\gamma_\lambda(1 + i\gamma_5)|\nu_\mu]J^\lambda + c \tag{8.13}$$

where J^λ is a hadronic current. Within the quark model, the commutation and conservation properties of J^λ allowed it to be identified with the quark current

$$J^\lambda = \bar{u}\gamma^\lambda(1 + i\gamma_5)d\cos\theta_c + \bar{u}\gamma(1 + i\gamma_5)s\sin\theta_c\ . \tag{8.14}$$

Here θ_c is an angle known as the Cabibbo [3] angle. Experiments on nuclear processes decaying from one state to another plus $e^+ + \nu_e$ and meson states decaying similarly, confirmed that G_F has almost the same value as that measured in the purely leptonic process $\mu^+ \rightarrow \bar{\nu} + e^+ + \nu_e$ and give for θ_c the value $\sin\theta_c = 0.220 \pm 0.003$. The natural conclusion was that the quarks provide another $SU(2) \times U(V)$ doublet with the form

$$\left(\frac{1+i\gamma_5}{2}\right)\begin{bmatrix} u \\ d\cos\theta_c + s\sin\theta_c \end{bmatrix}, \tag{8.15}$$

together with right-handed singlets adjusted to give the quark charges $\frac{2}{3}e$ and $(-)\frac{1}{3}e$. This has many problems, particularly yielding decay values for processes, such as $K^0 \rightarrow \mu^+ + \mu^-$, many orders of magnitude greater than observed. Eventually this situation was clarified by Glashaw, Iiopoulos, and Maiani [4] who proposed that there was an extra term in J^λ of the form

$$\bar{c}\gamma^\lambda(1 + i\gamma_5)[(-)d\sin\theta_c + s\cos\theta_c] \tag{8.16}$$

where c is a fourth quark with charge $\frac{2}{3}e$. The charged current may now be written as

$$J^\lambda = (\bar{u}\cos\theta_c - \bar{c}\sin\theta_c)\gamma^\lambda(1 + i\gamma_5)d + (\bar{u}\sin\theta_c + \bar{c}\cos\theta_c)\gamma^\lambda(1 + i\gamma_5)s, \tag{8.17}$$

which became known as the GIM mechanism and attracted large wagers. The main effect of this change was to suppress loop diagrams for $s + \bar{d} \rightarrow d + \bar{s}$, bringing the rate for $K^0 - \bar{K}^0$ oscillations in agreement with the experiment.

It was later noted by Weinberg [5] that this solves the problem of the strangeness changing Z^0 interactions. In the context of the $SU(2) \times U(1)$ gauge theory the combination $(-)d_L \sin\theta_c + s_L \cos\theta_c$ cannot be a singlet but must be part of another doublet

$$\left(\frac{1+i\gamma_5}{2}\right)\begin{bmatrix} c \\ (-)d\sin\theta_c + s\cos\theta_c \end{bmatrix}. \tag{8.18}$$

Particles containing the c quark in a $c - \bar{c}$ bound state were discovered in 1974 by Aubert et al. [6] and by Augustin et al. [7] and indicated a mass $m_c \approx 1.5G_a V_1$, which is not precise. This completed two generations of quarks and leptons.

The first sign of a third generation was the discovery of a third charged lepton by Perl et al. [8], the τ. Then a fifth quark type, the b (beauty), was discovered by Herb et al. [9], with charge $(-)\frac{1}{3}e$ and a mass of about $4.5G_eV$. A sixth quark type, the t top with charge $\frac{2}{3}e$, became theoretically necessary. It was eventually discovered by Ellis et al. [10], giving a combined value of the experimental results in the previous references of $181 \pm 12G_eV$ in 1995.

The real point of the Lagrangian formalism is that it provides a natural framework for the quantum mechanical implementation of symmetry principles. The dynamical equations in the Lagrangian formalism take the form of a variational principle, the principle of stationary action. Consider any infinitesimal transformation of the fields

$$\Psi^k(x) \to \Psi^k(x) + i\varepsilon f^k(x) \tag{8.19}$$

that leaves the action invariant

$$0 = \delta I = i\varepsilon \int d^4x \frac{\delta I[\Psi]}{\delta \Psi^k(x)} f^k(x) \,. \tag{8.20}$$

If ε is a constant, such symmetries are known as *global symmetries*. Of course, this is automatically satisfied for all infinitesimal variations of the fields if the fields satisfy the dynamical equations. By an infinitesimal symmetry transformation we mean one that leaves the action invariant even when the dynamical equations are not satisfied. If we now consider the same transformation with ε an arbitrary function of position in space–time we see that

$$\Psi^k(x) \to \Psi^k(x) + i\varepsilon(x) f^k(x) \tag{8.21}$$

then, in general, the variation of the action will not vanish, but it will have to be of the form

$$\delta I = (-) \int d^4x J^{\mu}(x) \frac{\partial \varepsilon(x)}{\partial x^{\mu}} \tag{8.22}$$

in order that it should vanish when $\varepsilon(x)$ is constant. If we now take the fields in $I[\Psi]$ to satisfy the field equations then I is stationary with respect to arbitrary field variations that vanish at forge spacetime distances, including variations of the form Equation (8.20), so in this case Equation (8.21) should vanish. Integrating by parts, we see that $J^{\mu}(x)$ must satisfy a conservation law

$$\frac{\partial J^{\mu}(x)}{\partial x^{\mu}} = 0 \,. \tag{8.23}$$

It follows immediately that

$$\frac{dF}{dt} = 0 \tag{8.24}$$

$$\text{where} \quad F \equiv \int d^3x J^0. \tag{8.25}$$

There is one such conserved J^{μ} and one constant of the motion F for each independent infinitesimal symmetry transformation. This represents a general feature of the canonical formalism, often referred to as Noether's theorem: *symmetries imply conservation laws*. Einstein was so impressed that he arranged for the German original to be translated into English and the final outcome is given in Reference [11]. Note that the English version, being much later, contains an outline of the applications that Noether's theorem has subsequently found, especially in the calculus of variations and relativity theory.

References

1. M. Gell-Mann, Cal. Tech. Synchroton Laboratory Report CTSL-20 (1961), unpublished. This was reproduced along with other articles on $SU(3)$ symmetry in M. Gell-Mann and V. Ne'eman, *The Eightfold Way*, Benjamin, New York, 1964.
2. K.J. Barnes, P.D. Jarvis, and I.J. Ketley, An orthogonal way: A synthesis for leptons and hadrons. *J. Phys.* G5 no. 1 (1979): 1.
3. N. Cabibbo, *Phys. Rev. Lett.* 10 (1963): 531.
4. S.L. Glashow, J. Iiopoulos, and L. Maiani, *Phys. Rev.* D2 (1970): 1285.
5. S. Weinberg, *Phys. Rev. Lett.* 27 (1971): 1688; *Phys. Rev.* 5 (1972): 1413.
6. J.J. Aubert et al., *Phys. Rev. Lett.* 33 (1974): 1404.
7. J.E. Augustin et al., *Phys. Rev. Lett.* 33 (1974): 1406.
8. M.L. Perl et al., *Phys. Rev. Lett.* 35 (1975): 1489.
9. S.W. Herb et al., *Phys. Rev. Lett.* 39 (1977): 252.
10. J. Ellis, G.L. Fogli, and E. Lisi, CERN-BARI preprint hep-pn /9507421.
11. E. Noether, *Nachr. Akad. Wiss. Götingen Math-Phys. Kl.* II (1918): 235. M.A. Tavel. *Transport Theory Statist. Phys.* 1, no. 3, (1971): 183.

Problems

8.1 Calculate the differential and total cross-sections for $e^+e^- \to \mu^+\mu^-$ in lowest order in e. The electron and meson spins are not observed.

8.2 Calculate the differential cross-section for electron–electron scattering to lowest order in e. Assume that final and initial spins are not measured.

8.3 Carefully defining your own notation, including normalization, derive the 3×3 matrix form of the Gell-Mann λ matrices.

8.4 State what you understand by the Cabibbo angle and give a rough numerical value for its sine.

8.5 Explain how the Cabibbo angle entered into the standard model, and say why this natural idea had very real problems.

8.6 Carefully explaining your notation, explain how Glashow, Iliopoulos, and Maiani made a proposal to remove the difficulties mentioned in Problems 8.5 and 8.6.

8.7 How did the Glashow, Iliopoulos, and Maiani, (GIM) mechanism first reveal its potential?

8.8 Say in your own words what Weinberg noted about the GIM mechanism, which revealed its possible potential.

8.9 What else did Weinberg notice about the GIM mechanism and how did he improve matters?

8.10 What did Perl et al. discover and what was its significance for model building?

8.11 Say in your own words what the real point is of the Lagrangian formalism for model builders.

8.12 What does your answer to Problem 8.11 lead to?

8.13 What is the familiar way of describing the content of Noether's theorem?

9

Spontaneous Symmetry Breaking and the Unification of the Electromagnetic and Weak Forces

The usual starting point for this topic is the wine bottle potential depicted in Figure 9.1. Clearly, there is an unstable position of equilibrium for a ball under "gravity" on top of the hump in the center. Once this is disturbed there is a spontaneous breaking of the symmetry and one position on the horizontal circle (shown dotted) is selected at random. Obviously we are really talking vacuum expectation values and states here. The dotted line around the lowest point of the potential is a set of massless states of equal energy, which can be transported with effectively zero force. However, the selected state has a mass, as we see by the fact that it takes energy to move it up the wall. In practice the "bottle" is represented by a quartic in scalar fields with a positive coefficient for the fourth power terms but a negative coefficient for the quadratic terms. So we can write

$$a(\phi^2)^2 - b\phi^2 \text{ with } a > 0 \text{ and } b < 0\,, \tag{9.1}$$

which has a minimum when

$$2a\phi^2 - b = 0\,, \tag{9.2}$$

$$\text{really} \quad <\phi^2> = \frac{b}{2a}\,, \tag{9.3}$$

where the lowest point is

$$a\frac{(b^2)^2}{4a^2} - \frac{b^2}{2a} = \frac{b^2}{4a}(b^2 - 2a), \tag{9.4}$$

which can be set to zero by putting $b = (-)\sqrt{2a}$. The potential is then

$$a(\phi^2)^2 - \sqrt{2a}\phi^2\,. \tag{9.5}$$

This is perhaps a good moment to mention the two volumes of *The Quantum Theory of Fields* by Nobel Prize winner S. Weinberg [1, 2], which are a meticulous and detailed presentation of the standard model of elementary particles by one of its greatest exponents. Indeed, if one is interested in physics beyond

DOI: 10.1201/9781439895207-9

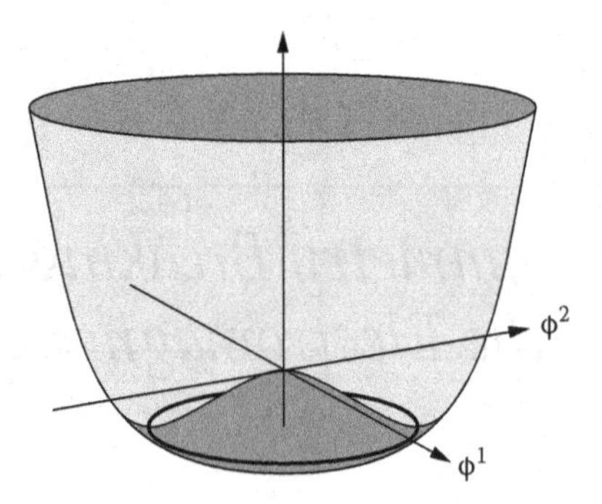

FIGURE 9.1
The wine bottle potential.

the standard model and the role of string theories in higher dimensions (which is the only known way to correctly introduce Einstein's gravity) then Weinberg's two following volumes [3,4] on these topics cannot be recommended too highly.

At any rate, we have adopted the notation of the first two volumes for this book, with the one exception that the parameters of group transformations carry opposite signs as a consequence of our choice of an active viewpoint.

With the charge on the proton taken as e, we have the change operator Q related to the hypercharge Y by

$$Q = T_3 + Y \qquad \text{and} \qquad Y = \frac{B + S - L}{2} \tag{9.6}$$

where B is baryon number, S is strangeness, and L is lepton number.

As a consequence, the up and down quarks that make up the charges of the proton and neutron, appearing as they do in triplets as a result of their colors being absorbed into singlets, have charges $\frac{2}{3}$ and $\frac{1}{3}$, respectively. Similarly the negative charged electron e^- and its associated neutral neutrino ν^0 form of doublet $\left(\begin{smallmatrix} e^- \\ \nu_e^0 \end{smallmatrix}\right)$ of $SU(2) \times U(1)$. As far as we know there are two further generations of this structure where the quarks are called charm and strange, then top and beauty (or bottom) in the third generation. Similarly the lepton pairs are the muon and its neutrino, then the tau and its neutrino form doublets at increasing masses of the charged particles in doublets

$$\begin{pmatrix} \mu^- \\ \nu_\mu^0 \end{pmatrix} \qquad \text{and} \qquad \begin{pmatrix} \tau^- \\ \nu_\tau^0 \end{pmatrix} \tag{9.7}$$

where the neutrinos were all thought to be massless. We say "were" because recent experiments have shown that at least one neutrino must have a mass that is not zero. The key experiments show oscillations between different neutrino types. At any rate the doublets of charged leptons and their associated neutral partners are left-handed ones projected by

$$\frac{1}{2}(1+\gamma_5)\,, \tag{9.8}$$

while the singlet positive right-handed charged partners of the charged leptons (positron, μ^+ and τ^+) are projected by

$$\frac{1}{2}(1-\gamma_5)\,. \tag{9.9}$$

Returning to the wine bottle potential, we introduce a doublet $Y = 1$ made from a positive and a neutral meson and denoted by $\binom{\phi^+}{\phi^0}$. With the potential written as

$$V = \mu^2\phi^+\phi + \lambda(\phi^+\phi)^2, \text{ with } \mu < 0 \text{ and } \lambda > 0, \tag{9.10}$$

the ϕ develops an expectation value

$$<\phi> = \frac{1}{\sqrt{2}}\begin{pmatrix}0\\ \nu\end{pmatrix}, \text{ with } \nu^2 = \frac{(-)\mu^2}{\lambda} + 0(\hbar)\,, \tag{9.11}$$

as a result of the spontaneous symmetry breaking from $SU(2)\times U(1)$ which is a combination of the obvious $U(0)$ factor with the $U(1)$ given by the unbroken $U(1)$ generated in the $SU(2)$ by τ_3.

Returning to the vector bosons (recall A^i_μ and $F^i_{\mu\nu}$) that carry the forces of electromagnetism and similarly the ones carrying the $U(1)$, which we now denote by B_μ and $F_{\mu\nu}$, we note that they must all be massless (to start with, at least until the symmetry is spontaneously broken to $U(1)$) because there is no invariant $A^i_\mu A^\mu_i$ (or $B_\mu B^\mu$) as the keen student can easily test. The Yang–Mills [5] Lagrangian is

$$\begin{aligned}(-)\frac{1}{4}(F^i_{\mu\nu})(F^{\mu\nu}_i) &- \frac{1}{4}(F_{\mu\nu}F^{\mu\nu})\\ &= (+)\frac{1}{4}(\partial_\mu A^i_\nu - \partial_\nu A^i_\mu + gA^j_\mu A^k_\nu \varepsilon^{ijk})^2 - \frac{1}{4}(\partial_\mu B_\nu - \partial_\nu B_\mu)^2,\end{aligned} \tag{9.12}$$

which follows from the definition

$$F^i_{\mu\nu} = \partial_\nu A^i_\mu - \partial_\mu A^i_\nu + f^{ijk}A^j_\mu A^k_\nu \tag{9.13}$$

by substituting $f^{ijk} = -ig\varepsilon^{ijk}$ and similarly 0 for the $U(1)$ case.

What we know is that after spontaneous symmetry breaking there is one vector boson field of charge e with mass m_W given by

$$W^\mu = \frac{1}{\sqrt{2}}(A^\mu_1 + iA^\mu_2) \tag{9.14}$$

and another of charge $(-)e$ with mass m_W given by

$$W^{\mu} = \frac{1}{\sqrt{2}}(A_1^{\mu} - i A_2^{\mu}) . \tag{9.15}$$

There are also two electrically neutral vector boson fields of masses m_Z and 0, respectively, given by orthonormal linear combinations of A_3^{μ} and B^{μ}. They are

$$Z^{\mu} = \cos(\theta) A_3^{\mu} + \sin(\theta) B^{\mu} \tag{9.16}$$

$$A^{\mu} = (-)\sin(\theta) A_3^{\mu} + \cos(\theta) B^{\mu} . \tag{9.17}$$

We have identified the A^{μ} as the massless photon field from knowing Q. This mixing angle is usually known as the Weinberg angle.

To complete this picture we write down a Yukawa [6] coupling of the lepton doublet to the charged scalars in the form

$$\mathcal{L}_{\phi e} = (-)G_e \overline{\begin{pmatrix} \nu_e \\ e^- \end{pmatrix}}_L \begin{pmatrix} \phi^+ \\ \phi^0 \end{pmatrix} e_R + h.c. . \tag{9.18}$$

It is possible that there are other scalar multiplets in the theory (and in extensions of the standard model, such as supersymmetry, they are compulsory), but we will not consider them here.

Clearly we need a gauge invariant term involving scalar and gauge fields. The most general form consistent with $SU(2)\times U(1)$ gauge invariance, Lorentz invariance (and, although we do not treat this here, consistent with renormalizability), is

$$\begin{aligned} \mathcal{L}_{\phi} = (-)\frac{1}{2} \left|(\partial_{\mu} - i A_{\mu}^i T^i(\phi) - i B_{\mu} y^{(\phi)} \phi\right|^2 \\ - \frac{\mu^2}{2}\phi^+\phi - \frac{\lambda}{4}(\phi^+\phi)^2 , \end{aligned} \tag{9.19}$$

where $\lambda > 0$. It is possible to perform an $SU(2) \times U(1)$ gauge transformation to a unitary gauge, in which $\phi^+ = 0$ and ϕ^0 is Hermitian, with a positive vacuum expectation value. The real part of ϕ^0 is the only physical scalar field.

The scalar Lagrangian then yields a vector meson mass term of the form

$$\begin{aligned} (-)\frac{1}{2}\left|\left(A_{\mu}^i T^{i(\phi)} + B_{\mu} y^{(\phi)}\right) < \phi >\right|^2 &= (-)\frac{1}{2}\left|\left(\frac{g}{2}A_{\mu}^i \tau^i - \frac{g'}{2}B_{\mu}\right)\begin{pmatrix} 0 \\ \nu \end{pmatrix}\right|^2 \\ &= (-)\frac{\nu^2 g^2}{4} W_{\mu}^{\dagger} W^{\mu} - \frac{\nu^2}{8}(g^2 + g'^2) Z_{\mu} Z^{\mu}. \end{aligned} \tag{9.20}$$

The photon mass term is zero as expected. The $W^{\pm}$ and the Z^0 have masses

$$m_W = \frac{\nu|g|}{2} \quad \text{and} \quad m_z = \frac{\nu\sqrt{g^2 + g'^2}}{2}, \tag{9.21}$$

respectively. Also, in the lowest order the electron now has a mass

$$m_e = G_e v \,. \tag{9.22}$$

The extension to include the other two generations of leptons is straightforward with the e and ν_e replaced by μ and ν_μ and τ and ν_τ, respectively. Also replace G_e by

$$G_\mu = G_e\left(\frac{m_\mu}{m_e}\right) \quad \text{and so forth.} \tag{9.23}$$

In the case of the muon, the exchange of W between low energy e and μ leptons produces an effective interaction

$$\left(\frac{g}{\sqrt{2}}\right)^2 \frac{1}{m_W^2}\left[\bar{e}\gamma^\lambda\left(\frac{1+\gamma_5}{2}\right)\nu_e\right]\left[\overline{\nu}_\mu\gamma_\lambda\left(\frac{1+\gamma_5}{2}\right)\mu\right] + h.c., \tag{9.24}$$

which may be compared to effective V-A theory, which gives

$$\frac{G_F}{\sqrt{2}}[\bar{e}\gamma^\lambda(1+\gamma_5)\nu_e][\overline{\nu}_\mu\gamma_\lambda(1+\gamma_5)\mu] + h.c., \tag{9.25}$$

where G_F is the conventional Fermi coupling constant, which is known from the good description of muon decay to have the value

$$G_F = 1.6663q(2) \times 10^{-5} GeV^{-2} \,. \tag{9.26}$$

Comparison yields

$$\frac{g^2}{m_W^2} = 4\sqrt{2}G_F \tag{9.27}$$

so that

$$v = \frac{2m_W}{g} = \frac{1}{2^{\frac{1}{4}}G_F^{\frac{1}{2}}} = 247\text{Gev.} \tag{9.28}$$

We also see that

$$G_e = \frac{0.511\text{Mev}}{247\text{GeV}} = 2.07 \times 10^{-6}, \tag{9.29}$$

which is a very small value. Notice that $m_Z > m_W$. In terms of the weak mixing angle

$$m_W = \frac{ev}{2|\sin(\theta)|} = \frac{37.3\text{GeV}}{|\sin(\theta)|}, \tag{9.30}$$

$$m_Z = \frac{ev}{2|\sin(\theta)||cos(\theta)|} = \frac{74.6\text{GeV}}{|\sin(2\theta)|} \,. \tag{9.31}$$

There are many corrections, of course. In particular there is a very large radiative correction. Also we know that the coupling constants change with energy scale. The best values are possibly

$$m_W = \frac{38.4\text{GeV}}{|\sin(\theta)|} \quad \text{and} \quad m_Z = \frac{76.9\text{GeV}}{|\sin(2\theta)|} . \tag{9.32}$$

It is now clear that, whatever the value of the Weinberg angle, these masses were too large to allow W or Z to be found before the early 1970s. Neutral current processes produced by Z^0 exchange gave perhaps the earliest confirmation of the theory in 1973 in a bubble chamber of $\nu_\mu - e^-$ elastic scattering. (As an interesting side note, the Nobel Prize shared by three authors [7] was awarded before the discovery of the $W^\pm$ and Z^0 and despite strong advice to the contrary.)

Moving on, the quark current

$$J^\lambda = \bar{u}\gamma^\lambda(1+\gamma_5)d\cos(\theta_c) + \bar{u}\gamma^\lambda(1+\gamma_5)_s d\sin(\theta_c) , \tag{9.33}$$

where θ_c is the Cabibbo [8] angle, had already been used in the 1960s to understand weak interactions between leptons and hadrons by the low-energy effective Lagrangian

$$G_F[\bar{e}\gamma_\lambda(1+\gamma_5)\nu_e + \bar{e}\gamma_\lambda(1+\gamma_5)\nu_\mu]J^\lambda + h.c.. \tag{9.34}$$

Experiments on processes such as $O^{14} \to N^{14\nu} + e^+ + \nu_e$ showed that G_F had much the usual value and that $\sin(\theta_c) = 0.220 \pm 0.003$. This led to other mixings and eventually to the hadronic current

$$J^\lambda = \overline{\begin{bmatrix} u \\ c \\ t \end{bmatrix}} \gamma^\lambda(1+\gamma_5)V \begin{bmatrix} d \\ s \\ b \end{bmatrix} \tag{9.35}$$

between all three generations of quarks, where V is a 3×3 unitary matrix named after Kobayashi and Maskawa [9]. Much interest in research today centers on this matrix, and the existence of three families allows a complex phase in the matrix and thus to T and CP violation. One of the best tests of this part of the standard model is the annihilation of e^+e^- in colliding beans. This reveals resonances (of then unexpected narrowness) and threshold steps to plateaux of calculable heights.

References

1. S. Weinberg, *The Quantum Theory of Fields. Volume I: Foundations.* Cambridge University Press, Cambridge, 1995.
2. S. Weinberg, *The Quantum Theory of Fields. Volume II: Modern Applications.* Cambridge University Press, Cambridge, 1996.
3. S. Weinberg, *Gravitation and Cosmology.* Wiley, New York, 1972.

4. S. Weinberg, *The Quantum Theory of Fields. Volume III: Supersymmetry*. Cambridge University Press, Cambridge, 2000.
5. C.N. Yang and R.L. Mills, *Phys. Rev.* 96 (1954): 191.
6. H. Yukawa, *Proc. Phys. Math. Soc. Japan* 17 (1935): 48.
7. S.L. Glashow, *Nucl. Phys.* 22 (1961): 579; A. Salam, *Elementary Particle Theory*, ed. N. Svartholm. Almqvist, Stockholm, 1968, p. 367; S. Weinberg, *Phys. Rev. Lett.* 19 (1967): 1264.
8. N. Cabibbo, *Phys. Rev. Lett.* 22 (1969): 156.
9. M. Kobayashi and K. Maskawa, *Prog. Theor. Phys.* 49 (1972): 282.

Problems

9.1 Go through the argument from Equation (9.1) to (9.5) and draw your own wine bottle potential.

9.2 Using Equation (9.6) work out your own changes for up and down quarks and for the charges of the electron and its associated neutrino.

9.3 Using the Weinberg angle, work out your own versions of Z^{μ} and A^{μ}.

9.4 Work out your own versions of m_w and m_z.

9.5 Show that a 2×2 unitary matrix has no complex phase but a 3×3 unitary matrix does have one complex phase.

10

The Goldstone Theorem and the Consequent Emergence of Nonlinearly Transforming Massless Goldstone Bosons

At the deepest level this chapter and Chapter 13 could be viewed as an introduction to the work of Einstein on general relativity where he used curvature of the three spatial and one time dimensions of the world in which we live to describe the theory of gravity. Any reader fortunate enough to have a strong background in general relativity can regard this book as light bedtime reading. I had such a fortunate background from my undergraduate days reading special mathematics at King's College London, then arguably the leading place in the world for gravitational research. My meeting with quantum mechanics terminated my possible research in the gravitational area with my heroes F. Pirani and H. Bondi. (My deep gratitude is to Pirani for helping me to change course to work with P. Matthews and A. Salam at Imperial College London.) At any rate, the idea is to use S_2, the two-dimensional surface of the points equidistant from the origin in three spatial dimensions, as the simplest nontrivial curved space as an introductory example, which we can all easily visualize, but to handle the mathematical details by the equivalent machinery to that used in more dimensions.

In the United Kingdom, most serious general relativity research is done in mathematics departments, and even when a physics or astronomy department has a serious research interest, the subject is rarely covered at undergraduate 3 or 4 levels. At Southampton University, where there is a strong general relativity group and even an appropriate book written in house, no astronomy or astrophysics students are taught general relativity nor even encouraged (forced?) to attend the one course in the mathematics 4th undergraduate year. Only rarely do dedicated theoretical high energy physics students take the course in their first year of research. The situation is often far worse at other institutions. The hope is that this chapter and Chapter 13 may act as a stepping stone for at least some of those students.

What exactly are Goldstone bosons? Whenever a simple Lie group is broken spontaneously (smoothly and in an arbitrary "direction") then a theorem by J. Goldstone [1] tells us that massless Goldstone bosons (scalar or pseudoscalar) are produced carrying the quantum numbers of the generator of the Lie group

DOI: 10.1201/9781439895207-10

along the broken "directions." We shall see later how these bosons interact with each other, why they are massless, and how they interact with other (standard field) states of the system. For the moment, we shall concentrate on the breaking of SU_2 to U_1. Here the SU_2 can be thought of as generated by rotations about the three spatial directions, and broken to leave simply the one generated by rotations about the third axis. Thus, the Goldstone bosons are associated with the first directions. If we call these bosons ϕ_1 and ϕ_2, they can be thought of as interpolating fields for the two Goldstone bosons (subject to appropriate boundary conditions we shall meet later) with the transformations between them under rotations (we take the active point of view) as corresponding changes of the fields. Alternatively, we can consider these ϕ_A, $A = 1, 2$, to be coordinates on the two-dimensional surface of the sphere and again (subject to conditions) to carry the information about the transformation of the points on the sphere.

With an eye to future developments, we can regard the surface of the sphere as a coset space manifold of dimension two. If we introduce Pauli matrices by

$$\sigma_i \text{ with } i = (1, 2, 3)$$

$$\text{with} \quad \sigma_1 = \begin{pmatrix} 0 & 1 \\ 1 & 0 \end{pmatrix} \qquad \sigma_2 = \begin{pmatrix} 0 & -i \\ i & 0 \end{pmatrix} \qquad \sigma_3 = \begin{pmatrix} 1 & 0 \\ 0 & -1 \end{pmatrix} \tag{10.1}$$

then the SU_2 generated can be represented by

$$Q_i = \exp\left(\frac{-i}{2}\sigma_i \xi_i\right) \tag{10.2}$$

where the ξ_i can be thought of as angles of rotation about the respective axes x, y, z on $(1, 2, 3)$, where $\xi^2 = \xi_i \xi_i$ is an invariant, with the consequence that we find the commutators of the generators are

$$[Q_i, Q_j] = i\epsilon_{ijk} Q_k \tag{10.3}$$

where summation is implied over the repeated index, and the Levi-Civita totally antisymmetric tensor is specified by

$$\begin{aligned} \epsilon_{ijk} &= 1 \text{ if } (ijk) \text{ are cyclic on } (1, 2, 3) \\ &= -1 \text{ if } (ijk) \text{ are anticyclic on } (1, 2, 3) \\ &= 0 \text{ if any pair of } (ijk) \text{ are equal.} \end{aligned} \tag{10.4}$$

Here we have used

$$\sigma_i \sigma_j = \delta_{ij} 1 + i\epsilon_{ijk} \sigma^k \tag{10.5}$$

$$\text{where} \quad 1 = \begin{pmatrix} 1 & 0 \\ 0 & 1 \end{pmatrix}. \tag{10.6}$$

The keen students are strongly advised to become familiar with the index notation and the Pauli matrices if they are not already competent in their use.

If we designate the coset space represented by the two-dimensional surface of the sphere, S_2, by

$$L = \exp\left(\frac{-i}{2}\xi_A \sigma_A\right) \tag{10.7}$$

where $A = (1, 2)$, then on left multiplication by a member of the U_1 subgroup with parametrization h_1 we get

$$h_1 L = h_1 L h_1^{-1} h_1 \tag{10.8}$$
$$= L' h_1 \tag{10.9}$$

and using

$$L'(\xi) = L(\xi') \tag{10.10}$$

reveals that

$$L(\xi') = h_1 L(\xi) h_1^{-1} \tag{10.11}$$

so that the action of the U_1 subgroup on the field in the coset space is simply that of rotation through the angle parametrizing the subgroup. Note that the completeness relation on Pauli matrices is simply

$$(\sigma_i)_{AB}(\sigma_i)_{CD} = 2\delta_{AD}\delta_{BC} - \delta_{CD}\delta_{AB} \tag{10.12}$$

when thought of from the matrix point of view.

Generalizing our $\frac{SU_2}{U_1}$ sphere to a more general coset space, we can write

$$gL = L'h \tag{10.13}$$

where g is the group (previously SU_2) and h is the subgroup (previously U_1), and L' is now a generalized version of $L = \frac{SU_3}{U_1}$. Clearly the transformations are nonlinear.

At this stage it is convenient to introduce the concept of the centralizer of a subgroup within a group. Here this would be written as $C_h(g)$. It simply means the collection of elements in g, which commute with h to give zero. In the S_2 case we had

$$C_{U_1}(SU_2) = U_1 \tag{10.14}$$

Now there is a very useful theorem by A. Borel [2] to the effect that when the centralizer is toroidal (a product of powers of U_1 factors) the coset space manifold is a Kähler manifold of order given by the sum of the powers of the U_1 factors. In the S_2 case, the order is just 1. Such a manifold can support supersymmetry of the same order as the order of the manifold. What does this mean and what are the implications? In the first place think about an ordinary two-dimensional plane with coordinates x and y. If you like, you can combine the coordinates into a complex variable, usually called z, by

$$z = x + iy \tag{10.15}$$

with conjugate $\bar{z}$ (or z^*) given by

$$\bar{z} = x - iy \tag{10.16}$$

and the question which at once arises is: Does the definition of the derivative of a single real variable apply to functions of a complex variable? The natural answer is that if $f(z)$ is a one-valued function, defined in a region of the Argand diagram, then $f(z)$ is differentiable at a point z_0 of that region. If $\frac{f(z)-f(z_0)}{z-z_0}$ tends to a unique limit as $z \to z_0$, provided z is also a point of that region, it is called the derivative of $f(z)$ at $z = z_0$ and is denoted by $f'(z_0)$.

A function of z which is one-valued and differentiable at every point of a domain D is said to be *holomorphic* (sometimes *equivalently regular* or *analytic*) in the domain D.

There is a necessary set of conditions for $f(z)$ to be holomorphic. If $f(z) = u(x, y) + iv(x, y)$ is differentiable at a given point z, the ratio of

$$\frac{(f(z) + \Delta z) - f(z)}{\Delta z}$$

must tend to a definite limit as $\Delta z \to 0$ in any manner. Now $\Delta z = \Delta x + i\Delta y$. Take Δz to be real, so that $\Delta y = 0$, then

$$\frac{u(x + \Delta x, y) - u(x, y)}{\Delta x} + i\left[\frac{v(z + \Delta x, y) - v(x, y)}{\Delta x}\right]$$

must tend to a definite limit as $\Delta x \to 0$. Clearly the partial derivatives $\frac{\partial u}{\partial x}$ and $\frac{\partial v}{\partial x}$ must exist at the point (x, y) and the limit is $\frac{\partial u}{\partial x} + i\frac{\partial v}{\partial x}$. Similarly, if we take $\Delta y = 0$ then $\frac{\partial u}{\partial y}$ and $\frac{\partial v}{\partial y}$ must exist at the point (x, y) and the limit is $\frac{\partial u}{\partial y} - i\frac{\partial v}{\partial y}$. Since the two limits must be identical, on equating real and imaginary parts we find

$$\frac{\partial u}{\partial x} = \frac{\partial v}{\partial y} \quad \text{and} \quad \frac{\partial u}{\partial y} = (-)\frac{\partial v}{\partial x}. \tag{10.17}$$

These two conditions are called the Cauchy–Riemann differential equations. Obviously, assuming differentiability is much stronger than assuming continuity.

The sufficient conditions for $f(z)$ to be holomorphic are that the continuous one-valued function $f(z)$ is holomorphic in a domain D if the four partial derivatives $\frac{\partial u}{\partial x}$, $\frac{\partial v}{\partial x}$, $\frac{\partial u}{\partial y}$, and $\frac{\partial v}{\partial y}$ are continuous and satisfy the Cauchy–Riemann equations at each point of D. Coordinates in the manifold are the Kähler coordinates. We shall stop this treatment here. Possibly we have already done too much. But this idea of holomorphy plays a huge role in many subjects.

References

1. J. Goldstone, *Nuovo Cimento* 9 (1961): 154.
2. A. Borel, *Proc. Natl. Acad. Sci. USA* 40 (1954): 147.

Problems

10.1 In your own words state what you understand about Goldstone bosons.

10.2 In your own words state what you understand regarding the surface of a sphere as a coset space of dimension two.

10.3 Define the Kronecker delta, δ_i, and the Levi-Civita tensor, ε_{ijk}.

10.4 Define the Pauli matrices σ_i with $i = (1, 2, 3)$ and the associated unit matrix 1.

10.5 Show that $\sigma_i \sigma_j = \delta_{ij}\,\mathbf{1} + i\varepsilon_{ijk}\sigma^k$.

10.6 Show how $SU(2)$ can be generated by a simple function of the Pauli matrices.

10.7 If $Q_i = \exp(-\frac{i}{2}\sigma_i \xi^i)$, calculate the commutator $[Q_i, Q_j]$.

10.8 State what you understand about the completeness relation and work this out for the case of the Pauli matrices.

10.9 If we designate the coset space represented by the two-dimensional surface of the sphere S_2 by $L = \exp(-\frac{i}{2}\xi_A \sigma^A)$ where $A = (1, 2)$, show that on left multiplication by a member of the U_1 subgroup with parameterization h_1 we get $h_1 L = L' h_1$ and find L'.

10.10 Using $L(\xi') = L'(\xi)$ find the action of the U_1 subgroup on the coset space and give your interpretation of this action.

10.11 Generalizing our $\frac{SU_2}{U_1}$ sphere to a more general coset space, $gL = L'h$ where g is the group, h is the subgroup, and L' is a generalized version of L. Do you think that the transformations are linear or nonlinear?

10.12 What is understood by the centralizer of a subgroup in a group, usually written as $C_h(g)$?

10.13 What does the Borel theorem tell us?

10.14 Combining the two-dimensional coordinates of a plane denoted by x and y into a single complex variable $z = x + iy$, state what you mean by the derivation of a function $f(z)$ at z_0.

10.15 State what you mean by a holomorphic function of z.

10.16 If $f(z) = u(x, y) + iv(x, y)$ show that the Cauchy–Riemann differential equations

$$\frac{\partial u}{\partial x} = \frac{\partial v}{\partial y} \quad \text{and} \quad \frac{\partial u}{\partial y} = (-)\frac{\partial v}{\partial x}$$

must hold. Is this differentiability weaker or stronger than continuity?

10.17 What are sufficient conditions for $f(z)$ to be holonomic? What are coordinates in the manifold to be known as?

11

The Higgs Mechanism and the Emergence of Mass from Spontaneously Broken Symmetries

There are actually several situations involving one, or more in the case of the supersymmetric version, Higgs boson. These can be presented in a variety of ways and in different gauges. We shall present one of these to make the basic idea as simple as possible. It is the first one studied by Peter Higgs [1] himself. We already know what happens when a gauge theory, such as the Goldstone model [2], is spontaneously broken. Massless scalars or pseudoscalars appear as Goldstone bosons corresponding to the broken generators. Then the massless gauge vector or pseudovector bosons absorb the Goldstone bosons and the result is that they develop masses. So now Higgs started by making the Goldstone model locally $U(1)$ invariant in the, by now, familiar way. We note from the start that we begin with four degrees of freedom, one each from the two scalars and two from the massless electromagnetic field potential $A^{\mu}(x)$ corresponding to the transverse waves of electromagnetism or equivalently the independent senses of polarization of the photons. To fully exhibit the particle interpretation of this theory, it is best to change to a "radial and phase" field description. Then we expect it to be simple when the symmetry becomes local, because it is exactly local phase variations that are being considered. In fact, the theory, now with A^{μ} present, is invariant under local phase changes. Indeed we can get rid of the phase field altogether. We start from

$$\phi = \frac{1}{\sqrt{2}}\left[f + \rho(x)\right]\exp\left[\frac{-i\theta(x)}{f}\right] \tag{11.1}$$

where $f\lambda = \mu$. Under a local phase (or gauge) transformation

$$\phi \to \exp[-ie\chi(x)]\phi(x), \tag{11.2}$$

$$A^{\mu} \to A^{\mu} + \partial_{\mu}\chi(x) \tag{11.3}$$

and we see that

$$\rho = \rho' \tag{11.4}$$

$$\theta' = \theta + ef\chi. \tag{11.5}$$

DOI: 10.1201/9781439895207-11

If we pick the particular gauge function χ by

$$\chi = (-)\frac{1}{ef}\,\theta, \tag{11.6}$$

we discover that $\theta' = 0$. So the phase of the gauge transformed ϕ can be made zero. But this is a locally varying phase, or a phase field, and its quanta are the massless Goldstone bosons of the spontaneously broken $U(1)$ symmetry. It seems that we can always pick a gauge χ such that at all points the phase field is zero. So there are no Goldstone bosons! The massless scalar or pseudoscalar particles have disappeared from the particle spectrum. But they correspond to the original scalar, or isoscalar, degrees of freedom. Consider what the original A^{μ} and ϕ have been replaced by now. We find that

$$A^{\mu\prime} = A^{\mu} - \frac{\partial^{\mu}\theta}{ef}, \tag{11.7}$$

$$\phi' = \frac{1}{\sqrt{2}}[f + \rho(x)]. \tag{11.8}$$

Consider the Euler–Lagrange. We find that for $A^{\mu\prime}$ we get

$$[\Box + e^2 f^2]A^{\mu\prime} - \partial^{\mu}\partial_{\nu}A^{\nu\prime} = (-)e^2 A^{\mu\prime}(\rho^2 + 2f\rho). \tag{11.9}$$

In the weak coupling, or perturbative, limit we can ignore the electromagnetic current on the right-hand side of this equation. The left-hand side is the force particle wave equation for a massive spin one particle. The operator M is replaced by $\Box + M^2$, where

$$M = ef. \tag{11.10}$$

A massive spin one particle has two transverse degrees of freedom together with one longitudinal, making three degrees of freedom. So we see what happened to the Goldstone degrees of freedom. They got "eaten" by the gauge field A^{μ} to make the massive gauge field

$$A^{\mu\prime} = A^{\mu} - \frac{\partial^{\mu}}{ef}. \tag{11.11}$$

The θ field is precisely the one whose quanta were Goldstone particles. This is the well known *Higgs mechanism*. In a sense, the two "masslessness" have canceled each other out. The result is that when a local $U(1)$ invariance is spontaneously broken there are no Goldstone particles but the massless $U(1)$ gauge field gains mass. There can be more than one Higgs boson; indeed with supersymmetric formulation there must be at least two. They can also be viewed in different gauges but they all come down eventually to what you have seen. Not only has supersymmetry not been found experimentally but neither has any Higgs boson. Notice that the Higgs boson does not specify its own mass. I am one of a growing group of theoreticians who believe that the Higgs boson will never be found in the sense it is understood here. I personally have started to look for an alternative structure but if I find one I shall still call it the Higgs boson.

References

1. P. W. Higgs, *Phys. Lett.* 12 (1964): 132.
2. J. Goldstone, *Nuovo Cimento* 9 (1961): 154.

Problems

11.1 Read through from Equation (11.1) to Equation (11.6). Now close the book and repeat the calculations in your own notation if you prefer.

11.2 Read through from Equation (11.7) to Equation (11.10). Now close the book and repeat the calculations in your own notation if you prefer.

11.3 Explain in your own words how to count the degrees of freedom after Equation (11.10).

11.4 Explain in your own words what is usually understood by the Higgs mechanism.

12

Lie Group Techniques for beyond the Standard Model Lie Groups

We may well start with technicolor, a scheme proposed by L. Susskind [1], which was based on a generalization of the color group $SU(3)$ of the strong interactions. It was briefly fashionable but was completely ruled out by experimental results and has now been forgotten by most serious researchers. The grand unified groups (or GUTs as they are often called) are the main source of possibilities. This starts historically with a Pati and Salam scheme [2]. Salam himself was the first to point out the two main problems. In all such schemes the proton becomes unstable and there is no sign of this experimentally. Also, all such theories must contain a magnetic monopole [3,4] and again there is no credible experimental candidate. By far the most fashionable candidate was the proposal by H. Georgi and S.L. Glashow [5] to use $SU(5)$, which contains $SU(3) \times SU(2) \times U(1)$ of the standard model and has a five-dimensional multiplet containing:

$$(3, \mathbf{1})_{-\frac{1}{3}} + (\mathbf{1}, \mathbf{2})_{\frac{1}{2}} \tag{12.1}$$

where the first factor in the brackets is the $SU(3)$ multiplicity, the second factor is the $SU(2)$ multiplicity, and the subscript is the $U(1)$ number normalized to give zero for this **5** representation of $SU(5)$. The various techniques we studied for the standard model Lie groups apply directly, including the classification theorem by Dynkin [6].

As we move on, an important constraint is that the groups we consider must contain the standard model groups. There are only four classes of simple groups, $SU(2n)$, $SU(2n-1)$, $SO(2n)$, and $SO(2n+1)$ where n is a positive integer. The symplectic groups $Sp(2n)$ seem never to have played a part.

The orthogonal groups are interesting because, like the unitary groups, they can have complex representations. As we shall see, $SU(5)$ can be easily embedded in an orthogonal group. The group $SO(10)$ [8], which has rank 5, contains the subgroup $SU(5) \times U(1)$ in two different ways. A spinor representation of $SO(10)$ is 16 dimensional and contains the $\mathbf{10} + \overline{\mathbf{5}} + \mathbf{1}$ under $SU(5)$. Hence, it neatly contains a single fermion family, plus a right-handed neutrino state. Higher dimensional orthogonal states contain several such families in a single irreducible representation. For $n \neq 6$ orthogonal groups are anomaly

DOI: 10.1201/9781439895207-12

free and thus explain the mystery of cancellations of the anomalies in the $\overline{\mathbf{5}}$ and $\mathbf{10}$ representations. This is all so very convenient that it is hard to see grand unification stopping at $SU(5)$.

We turn next to $SO(10)$, which is of rank 5. There are $\mathbf{45}$ gauge bosons transforming as the adjoint representation under $SU(5) \times U(1)$; the $\mathbf{45}$ of $SO(10)$ has the decomposition

$$\mathbf{45} = \mathbf{24}_0 + \mathbf{1}_0 + \mathbf{10}_1 + \overline{\mathbf{10}}_{-1}, \tag{12.2}$$

whereas the spinorial representations are $2^4 = 16$ dimensional. These are written as the product of five $SU(2)$ spinors. Now we wish to identify the fermion states. This is easily done by choosing the $SU(3)$ to be a subgroup of $SO(6)$ acting on the first three operator components and the $SU(2)$ to be a subgroup of the $SO(4)$ acting on the last two spinor indices. Clearly

$$SO(10) \supset SO(6) \times SO(4) \tag{12.3}$$

so the product structure of the subgroups $SU(3) \times SU(2)$ is ensured [8].

By construction, the $SU(5)$ gauge bosons couple to fermions, through the decomposition of

$$SO(10) \supset SO(6) \times SO(4) \tag{12.4}$$
$$= SU(4) \times SU(2) \times SU(2) \tag{12.5}$$
$$= SU(3)_{color} \times U(1) \times SU(2)_L \times SU(3)_R. \tag{12.6}$$

Under $SU(3)_{color} \times U(1) \times SU(2)_L \times SU(3)_R$ the fermions in the $\mathbf{16}$ transform as

$$\mathbf{16} = (\overline{\mathbf{3}}, \mathbf{1}, \mathbf{2}) + (\mathbf{1}, \mathbf{1}, \mathbf{2}) \tag{12.7}$$

and the $\mathbf{45}$ gauge bosons transform as

$$\begin{aligned}\mathbf{45} = {} & (\mathbf{8}, \mathbf{1}, \mathbf{1}) + (\mathbf{1}, \mathbf{3}, \mathbf{1}) + (\mathbf{1}, \mathbf{1}, \mathbf{3}) + (\mathbf{1}, \mathbf{1}, \mathbf{1}) \\ & + (\overline{\mathbf{3}}, \mathbf{2}, \mathbf{2}) + (\mathbf{3}, \mathbf{2}, \mathbf{2}) + (\mathbf{3}, \mathbf{1}, \mathbf{1}) + (\overline{\mathbf{3}}, \mathbf{1}, \mathbf{1}).\end{aligned} \tag{12.8}$$

It should be noted that only one of the two embeddings of the standard model $SU(3) \times SU(2) \times U(1)$ can be achieved as the spinor states in the other one are never singlets under $SU(2)$.

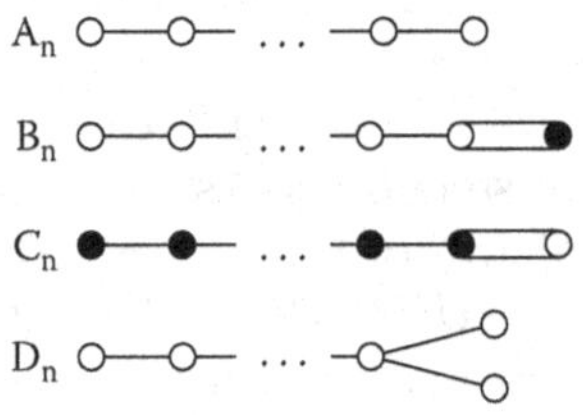

A_n

B_n

C_n

D_n

FIGURE 12.1
The multiplicities of the spin multiplets.

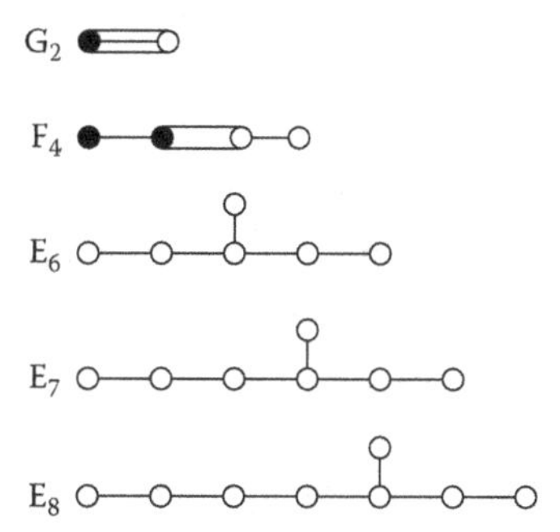

FIGURE 12.2
The four infinite families of the Dynkin classification.

The Lie group techniques we need have all been done in the standard model section or are trivial extensions. But it is important to record the Dynkin classification. There are four infinite families denoted in Figure 12.1 with the shorter vectors indicated by filled circles. Moving to the exceptional cases in increasing size we have of these, E_6 has the curious link to octonians which is the longest number base in which reality, commutation, and association have all been given up. Again, E_7 and E_8 are real, which effectively rules out light particles. There are a number of coincidences. First A_1, B_1, C_1 are all $SU(2)$. Second, $B_2 = C_2$. Third, $D_3 = A_3$. Fourth, if we remove one circle from D_3 to get D_2 it falls apart into two disconnected circles (the middle one must be removed to stay in the D_n family). Thus, D_2 is not simple. This is the important statement that the algebra of $SO(4)$ is the same as the algebra of $SU(2) \times SU(2)$. This is the complete list of such coincidences.

References

1. L. Susskind, *Phys. Rev.* D19 (1979): 2619.
2. J. Pati and A. Salam, *Phys. Rev. Letters* 31 (1973): 277.
3. G. 'tHooft, *Nucl. Phys.* B79 (1974): 276.
4. A.M. Polyakov, *JETP Letters* 20 (1974): 194.
5. H. Georgi and S.L. Glashow, *Phys. Rev. Letters* 28 (1972): 1494.
6. E.B. Dynkin, *Amer. Maths. Soc. Transl.* Series 2, 6 (1957): 111.
7. H. Georgi, *Lie Algebras in Particle Physics*, 2nd ed. (Frontiers in Physics, Vol. 54), Westview Press, Boulder, CO, 1999.
8. H. Georgi, in *Particles and Fields*, ed. C. Carlson, American Institute of Physics, New York, 1975.

Problems

12.1 What two main problems did Salam point out with GUT models?

12.2 Derive $\mathbf{5} = (\mathbf{3}, \mathbf{1})_{-\frac{1}{2}} + (\mathbf{1}, \mathbf{2})_{\frac{1}{2}}$ of Georgi and Glashow for $SU(5) \supset SU(3) \times SU(2) \times U(1)$.

12.3 Find, or look up, the two ways that $SU(5) \times U(1)$ can be embedded in $SO(10)$.

12.4 Show that the 16 of $SO(10)$ has the decomposition $\mathbf{10} + \overline{\mathbf{5}} + \mathbf{1}$ under $SU(5)$.

12.5 Show that the adjoint **45** representation of $SO(10)$ has the decomposition $\mathbf{45} = \mathbf{24}_0 + \mathbf{1}_0 + \mathbf{10}_1 + \overline{\mathbf{10}}_{-1}$ under the $SU(5) \times U(1)$.

12.6 How many components have the spinor representations of $SO(10)$?

12.7 Show how $SO(10) \supset SU(3)_{color} \times U(1) \times SU(2)_L \times SU(3)_R$.

12.8 Confirm the decomposition of the **16** in Equation (12.7).

12.9 Confirm the decomposition of the **45** in Equation (12.8).

12.10 Explain why the exceptional groups E_7 and E_8 are not thought to be useful in GUT schemes.

13

The Simple Sphere

Now what is supersymmetry? At the most basic level it is a symmetry that relates bosons (integer spin particles) to fermions (odd half-integer spins) in a way we shall examine in more detail in later chapters. There is no real evidence for the existence of this symmetry. However, it does seem to have almost magical powers in controlling infinities and improving the convergence of the $SU(3)$, $SU(2)$, and $U(1)$ running coupling constants at what is then regarded as a supersymmetric unification scale.

We have already seen that Kähler manifolds have the hyperkähler nature fixed by an integer and now we can reveal that this same integer is the one we met in the Borel theorem and is the centralizer of the denomination in the numerator when the coset space has the $\frac{G}{H}$ structure. We see that the Goldstone bosons are a pair of pseudoscalar mesons with equal but opposite charges $\pm e$ and they are massless in this approximation. The Kähler nature of the manifold ensures that each of the Goldstone bosons (pions in elementary particle physics) has a spin one-half partner of the same charge and also massless. You can see how very restricted the particle spectrum is in this model. Of course, the couplings are also specified so the scheme overall is very specified indeed.

Of course, it is just a model and the real world is probably totally different. (We do not know. Nobody has found supersymmetry experimentally yet.) Nevertheless there is a huge amount of literature in particle physics on this type of work. Much of it comes from foreign research groups with the contribution from Japan being both extensive and very good in mathematical terms. So if supersymmetry is ever discovered, there are many ingenious models to be tested.

To help us understand this section we shall think in more physical terms, where the $\frac{SU_1}{U_1}$ structure is embedded directly into the chiral σ-model as presented by P. Chang and F. Gürsey [1] and S. Weinberg [2, 3], although we shall retain our own familiar notation for the most part. This structure was much used in the 1960s with three pions (pseudoscalars) and one σ particle (scalar at the basic level), where the changed pions form an isotopic spin doublet and the σ is a singlet of isospin. This scheme did much to mirror current algebra results and in its nonlinear form gave a basis for a much used perturbative

DOI: 10.1201/9781439895207-13

scheme. The transformation laws in this scheme are

$$\pi_A \to \pi_A + \varepsilon_{A3B}\theta_3\pi_B + \phi_A\sigma \tag{13.1}$$

$$\sigma \to \sigma - \phi_A\pi_A \tag{13.2}$$

and we recognize a normal linear representation where the σ is a singlet. (Now in chiral $SU(2)\times SU(2)$ the corresponding multiplet is four dimensional; there are three pseudoscalar fields.) In this case the nonlinearity results from imposing the chiral $SU(2)$ invariant constraint

$$\sigma^2 + \pi_A\pi_A = f_\pi^2 \tag{13.3}$$

where f_π is a constant to be determined from experiment, to eliminate the unphysical σ field. The transformation law is then

$$\pi_A \to \pi_A + \varepsilon_{A3B}\theta_3\pi_B + \phi_B\left[f_\pi^2 - \pi^2\right]^{\frac{1}{2}}\delta_{AB} \tag{13.4}$$

where $\pi^2 = \pi_A\pi_A$ and we have arbitrarily selected the positive square root. We emphasize that this is an example of the transformation laws derived earlier for a particular choice of our arbitrary invariant θ. The simple form of this transformation law, together with the intuitive feeling for the nonlinearity arising from the constraint, have made this a popular choice in the literature. However, we now turn to the stereographic choice of coordinates used by B. Zumino [4] to allow the introduction of supersymmetry by emphasizing the Kähler properties of the 2-sphere. Things now look very different. The two real coordinates on the sphere (pseudoscalar fields) are replaced by a single complex variable z:

$$g_{z\bar{z}} = \frac{\partial^2 V}{\partial z \partial \bar{z}} \tag{13.5}$$

where V is a potential function so that the usual cross-derivative constraints are satisfied, for this to be a Kähler manifold. This is often referred to as the existence of an almost complex structure. In this framework, the nonlinear transformation law for the coordinates takes the form

$$z \to z + i\theta_3 z + \frac{c\omega}{2} + \frac{\overline{\omega}z^2}{2c} \tag{13.6}$$

where

$$\omega = \phi_1 + i\phi_2. \tag{13.7}$$

c is a constant (identifiable as $2f_\pi$) and we note that this transformation law is holomorphic in z. It is simple to change to a more familiar pair of real variables, now $x_A (A = 1, 2)$, by setting

$$z = x_1 + ix_2 \tag{13.8}$$

and we find that Equation (13.7) yields

$$x_A \to x_A + \varepsilon_{A3B}\theta_3 x_B + \phi_B \left[\frac{\delta_{AB}(c^2 - x^2)}{2c} + \frac{x_A x_B}{c} \right] \tag{13.9}$$

where

$$x^2 = x_A x_A \tag{13.10}$$

and Equation (13.7) has been used. It can be shown that [5]

$$\pi \cot(\theta) = \left[f_\pi^2 - \pi^2 \right]^{\frac{1}{2}}, \tag{13.11}$$

$$2cx \cot(\theta) = c^2 - x^2. \tag{13.12}$$

Direct comparison of these last two results gives

$$\pi^2 = \frac{4c^2 f_\pi^2 x^2}{[c^2 + x^2]^2}, \tag{13.13}$$

or equivalently

$$\pi_A = \frac{2c f_\pi x_A}{c^2 + x^2} \tag{13.14}$$

as the connection between the two coordinate systems. Note again that c may be identified with $2f_\pi$ when the equality of the two coordinate systems is transparent in the small field limit. I hope that this simple example makes clear the advantage of working with the general coordinate treatment whenever possible.

We start this long section by reviewing K.J. Barnes, P.H. Dondi, and S. Sarkar [6] and the structure of chiral $SU(2) \times SU(2)$ to establish notation. The transformation of the fundamental (quark) multiplet is specified by

$$q \to q - i\zeta_i \frac{1}{2}\sigma_i q - i\xi_i \frac{1}{2}\sigma^i (i\gamma_5) q \tag{13.15}$$

to the lowest order in the real parameters ζ_i and ξ_i, $i = 1, 2, 3$, where σ^i are the familiar Pauli matrices. Note the extra $i\gamma_5$ factor in the final term, which is included to ensure that the Goldstone bosons of this scheme will be pseudoscalar. We emphasize that this is precisely the usual symmetry of the quark and gluon QCD Lagrangian in the two flavor case (ignoring $U(1)$ complications), which leads to the familiar low-energy approximation to hadronic physics [7–10]. Our γ_5 is not Hermitian, but self-barred, so that under the transformations in Equation 13.15 the quark mass term $m\overline{q}q$ is not invariant and so should not appear in the unbroken Lagrangian, whereas the kinetic term proportional to $\overline{q}\gamma^\mu \partial_\mu q$ is invariant because the γ^μ anticommutes with the γ_5 in the axial generators. The crucial step in describing the Goldstone bosons is to parametrize the coset space defined by the quotient of the $SU(2) \times SU(2)$

by the vector $SU(2)$ parametrized by the ξ^i alone. This takes the simple form

$$\hat{L} = \exp\left\{-\frac{1}{2}i\,\xi n_i \sigma^i (i\gamma_5)\right\} \tag{13.16}$$

where the Goldstone fields are described by

$$M^i = Mn^i \tag{13.17}$$

with

$$(n^i)(n^i) = 1 \tag{13.18}$$

so that

$$(M^i)(M^i) = M^2 \tag{13.19}$$

and ξ is an arbitrary dimensionless function of the quotient of M by a constant f_π. Provided that ξ is proportional to this quotient in the limit of small fields, then f_π is proportional to the pion decay constant. This arbitrariness may be viewed as the freedom to change coordinate systems on the coset space or to redefine the field variables describing the mesons. Notice that the Goldstone fields M^i really do serve to describe three pseudoscalar pions as usual. This notation is reserved for this general coordinate system (as opposed to π^i for the nonlinear σ-model coordinates, say), and we stress again that if ξ is an arbitrary function of M (normalized to M for small fields) then all coordinate systems (with overlapping coordinate patches, i.e., not prohibited by singularities) are incorporated in this one description. If we define projection operators by

$$P_L = \frac{1}{2}(1 + i\gamma_5) \tag{13.20}$$

$$P_R = \frac{1}{2}(1 - i\gamma_5) \tag{13.21}$$

$$P_L P_L = P_L \tag{13.22}$$

$$P_R P_R = P_R \tag{13.23}$$

$$P_L P_R = 0 = P_R P_L \tag{13.24}$$

$$P_L + P_R = 1 \tag{13.25}$$

then we can write Equation (13.15) as

$$\hat{L} = L P_L + L^{-1} P_R \tag{13.26}$$

where L is unitary and the γ_5 dependence is now contained solely in the projection operators. It is then clear that we can deal with

$$L = \exp\left\{-\frac{1}{2}i\,\xi n_i \sigma^i\right\} \tag{13.27}$$

and reinstate the γ_5 factors only when wishing to consider the explicit couplings of the Goldstone bosons to matter fields. The action of a group element g (of $SU(2) \times SU(2)$) on the coset space can be specified by T. Clark and U.T. Veldhuis [11]

$$gL = L'h \tag{13.28}$$

where

$$L'(M_i) = L(M_i') \tag{13.29}$$

specifies the nonlinear transformations of the Goldstone boson fields,

$$h = \exp\left\{-\frac{1}{2}i\ \lambda_i \sigma_i\right\} \tag{13.30}$$

and the λ_i depend on the fields and the group parameters. What we have are nonlinear transformations among the M_i (which give a realization of the group), which are linear under the action of the $SU(2)$ subgroup, thus neatly describing a situation where the full group is still realized, but in a manner well suited to spontaneous breaking to the subgroup. The Goldstone bosons are a linear representation of the $SU(2)$ subgroup only. Although the procedure extends to other representations, for our purposes it will be sufficient to stay mostly in the fundamental representation.

We are now ready to discuss the chiral $SU(2)$ structure embedded in this framework. Consider the subgroup of the chiral $SU(2) \times SU(2)$ group specified in Equation 13.15 by retaining only the parameters ξ_3 and ξ_A with $A = 1$ and 2. Obviously this is an $SU(2)$ subgroup and we call it chiral $SU(2)$ in recognition of the $(i\gamma_5)$ factors with the σ^A generators. Clearly the σ^3 generates a $U(1)$ subgroup so that the coset space obtained by the quotient of chiral $SU(2)$ by this $U(1)$ is parametrized by coordinates M_A, $A = 1$ and 2, which can be viewed as describing two Goldstone pseudoscalars. Notice that the embedding of this $SU(2)/U(1)$ structure in the $\frac{SU(2)\times SU(2)}{SU(2)}$ structure is uniquely specified. Moreover, if we set M_3 and n_3 to zero in our previous discussion, then

$$L = \exp\left\{-\frac{1}{2}i\ \xi n_A \sigma^A\right\} \tag{13.31}$$

and set $\lambda_A = 0$ so

$$h = \exp\left\{-\frac{1}{2}i\ \lambda_3 \sigma^3\right\} \tag{13.32}$$

where ξ is now an arbitrary function of

$$M^2 = M_1^2 + M_2^2, \tag{13.33}$$

which when M_3 becomes zero remains as the only independent scalar. Although many readers will instantly appreciate the nature of this embedding, experience has taught us that confusion often arises at this point and it is hoped that a more detailed discussion will not divert readers too far from the real theme. Suppose in this quark model, we define the vector and axial currents, as usual, by

$$V_i^\mu = \bar{q}\gamma^\mu \frac{1}{2}\sigma_i q \qquad \text{and} \qquad A_i^\mu = \bar{q}\gamma^\mu \frac{1}{2}\sigma_i (i\gamma_5) q \tag{13.34}$$

and implement the transformations in Equation (13.17) by charges

$$Q_i^V = \int V_i^0 d^3x \qquad \text{and} \qquad Q_i^A = \int A_i^0 d^3x \tag{13.35}$$

by using free field communication relations. Naturally, while the symmetry is unbroken, the charges are time independent as a result of being constructed from the time components of the conserved Noether's currents. The chiral $SU(2) \times SU(2)$ can now be written as

$$[Q_i^V, Q_j^V] = i\varepsilon_{ijk} Q_k^V, \tag{13.36}$$

which is the algebra of the central (vector) subgroup, together with

$$[Q_i^V, Q_j^A] = i\varepsilon_{ijk} Q_k^A \tag{13.37}$$

confirming that the axial charges are in a three-dimensional representation of the vector subalgebra and

$$[Q_i^A, Q_j^A] = i\varepsilon_{ijk} Q_k^V \tag{13.38}$$

showing the closure of the axial parts of the algebra into the vector subalgebra and revealing the symmetric space structure which clarifies the coset space construction we introduced earlier. Now where is the chiral $SU(2)$ algebra embedded? As there are only two inequivalent types of $SU(2)$ in $SU(2) \times SU(2)$, this one must be equivalent to one of the more obvious ones. The left and right $SU(2)$ algebras defined by the generators

$$Q_i^L = \frac{1}{2}(Q_i^V + Q_i^A) \tag{13.39}$$

$$Q_i^R = \frac{1}{2}(Q_i^V - Q_i^A) \tag{13.40}$$

have the property that

$$[Q_i^L, Q_j^R] = 0 \tag{13.41}$$

so that the centralizer of either of these $SU(2)$ algebras in the $SU(2) \times SU(2)$ is the other. This is quite unlike the way in which the vector subgroup is embedded as seen in Equations (13.37) and (13.38) so that the chiral $SU(2)$ must be equivalent to the vector subgroup. We can take the unitary operator that implements this equivalence to be

$$U = \exp\left(\frac{1}{2} i\pi P_L \sigma^3\right), \tag{13.42}$$

which introduces the mapping

$$\begin{pmatrix} V_1 \\ V_2 \\ V_3 \end{pmatrix} \to \begin{pmatrix} A_1 \\ A_2 \\ A_3 \end{pmatrix}, \tag{13.43}$$

which is a trivial relabelling of the form we took with ξ_3 and ζ_A as parameters, and obviously has the correct commutation properties. This equivalence confirms that the coset space identified as the quotient of chiral $SU(2)$ by the $U(1)$ generated by V_3 is indeed the two-dimensional surface of a sphere as we have claimed. The mapping in Equation 13.43 clearly mixes parity types, so for the physical applications we have in mind the basis of σ_3 with $\sigma_A(i\gamma_5)$ as generators is the appropriate one justifying as it does our notation M_A as two fields describing pseudoscalar mesons, and dictating the form of the couplings to matter fields correspondingly, exactly as in the full $SU(2) \times SU(2)$ scheme. Note particularly that the chiral $SU(2)$ never appears as the denominator of the quotient defining a coset space. It is a subgroup of $SU(2) \times SU(2)$, which is equivalent to the vector $SU(2)$ but at no stage is considered as a conserved subgroup in a broken symmetry scenario. Thus the M_i and subsequently the M_A are always interpreted as fields describing pseudoscalar Goldstone bosons. The point is that chiral $SU(2)$ is a subgroup of chiral $SU(2) \times SU(2)$ and when the latter is spontaneously broken to the vector $SU(2)$ (with pseudoscalars M_i) then the chiral $SU(2)$ is broken to the $U(1)$ generated by V_3 (with M_A as the pseudoscalars). The broken chiral $SU(2)$ scheme (embedded uniquely in all possible broken chiral $SU(2) \times SU(2)$ supersymmetrizations) is unambiguously defined in the framework provided by the very simple Kähler structure of the 2-sphere [12].

We can now see the advantages of using this chiral 2-sphere as a model. It is simpler than the chiral $SU(2) \times SU(2)$ scheme, even in the purely bosonic sector. Moreover, the 2-sphere is a Kähler manifold and so admits a supersymmetric extension in which the Goldstone bosons acquire fermionic (Weyl) partners without yet more quasi-Goldstone bosons and fermions being forced into the model. Also the resulting couplings among particles are uniquely specified. Contrast this with the situations in References 14 and 15 where the number of bosons doubles as does the number of associated fermions, and finally the couplings involving these new particles are not uniquely specified. Of course, these latter cases are closer to the physics of the real world (they have three pions, for example) but the embedded chiral 2-sphere model retains many significant features and is a far more tractable theoretical laboratory.

In the following we give a treatment of most of the basic mathematics at the root of the subject of this book. Curiously, this begins by consideration of the embedding of the structure needed for the current problem into that of a larger system that has previously been solved in general coordinates leading to a closed form involving only simple functions [6]. The embedding is unique. Thus the starting point is a review of this established larger system and its solution, in which the liberty of changing notation (slightly) for convenience has been taken.

We recall that the $SU(2) \times SU(2)$ structure of (13.15), which we can conveniently rewrite as

$$q \to q - i\zeta_i \frac{\sigma^i}{2} q - i\xi_i \frac{\sigma^i}{2}(i\gamma_5)q \tag{13.44}$$

contains our S_2 coset space in the structure

$$\frac{[(-)\frac{1}{2}\zeta_1\sigma^i(i\gamma_5)][(-)\frac{1}{2}\zeta_2\sigma^2(i\gamma_5)]}{[(-)\frac{1}{2}\xi^3\sigma^3]} \tag{13.45}$$

as we noted previously. The $SU(2) \times SU(2)$ structure is spanned by the two sets of three orthogonal elements L_i and R_i satisfying the commutation relations

$$[L_i, L_j] = i\varepsilon_{ijk} L_k \tag{13.46}$$

$$[R_i, R_j] = i\varepsilon_{ijk} R_k \tag{13.47}$$

$$[L_i, R_j = 0] \tag{13.48}$$

and the linear combinations

$$V_i = L_i + R_i \tag{13.49}$$

$$A_i = L_i - R_i \tag{13.50}$$

are frequently used. The V_i generate an $SU(2)$ subgroup, which is parity conserving. An element of the $SU(2) \times SU(2)$ group may be specified by two sets of three real parameters and the alternative expressions

$$g = \exp(-i[\xi_i V_i + \zeta_i A_i]) \tag{13.51}$$

$$g = \exp\left(-i\theta_i^L L_i\right)\exp\left(-i\theta_i^R R_i\right) \tag{13.52}$$

will prove useful with

$$\theta_i^L = \xi_i + \zeta_i \tag{13.53}$$

$$\theta_i^R = \xi_i - \zeta_i \tag{13.54}$$

specifying the correspondence.

Every element of the group can be decomposed into a product of the form

$$g = \exp(-i\eta_i A_i)\exp(-i\eta_i V_i), \tag{13.55}$$

which is unique in a neighborhood of the identity element and this plays a crucial role in the general nonlinear realization scheme. The linear transformation laws are best specified by giving the quarks a Dirac spinor in the usual manner and taking

$$q \to q - \frac{i}{2}\theta_i^L \sigma^i \frac{1+\gamma_5}{2} q - \frac{i}{2}\theta_i^R \sigma^i \frac{1-\gamma_5}{2} q \tag{13.56}$$

as the concrete infinitesimal form.

Since the matrices

$$P_L = \frac{(1+\gamma_5)}{2} \tag{13.57}$$

$$P_R = \frac{1-\gamma_5}{2} \tag{13.58}$$

act as a standard set of projection operators, the treatment of linear transforming multiplets of $SU(2) \times SU(2)$ now follows trivially.

To treat the nonlinear realizations of $SU(2) \times SU(2)$ in full generality, the M_i of the adjoint vector of $SU(2)$ must be considered in more detail. In the terminology of L. Michel and L.A. Radicati [13], the vector is said to be generic (or belong to the generic structure) if all eigenvalues of M are distinct. For the generic case, the minimal polynomial for the matrix is the characteristic polynomial satisfying the equation

$$\prod_{A=1}^{3}(M - m_A) = 0 \tag{13.59}$$

where the m_A are the eigenvalues, which satisfy

$$\sum_{A=1}^{3} m_A = 0 \tag{13.60}$$

if the matrix is traceless. Thus the two vectors with components given by powers of the matrix in the form

$$M_i^\alpha = \frac{1}{2}\,\mathrm{Tr}\big([M]^\alpha \sigma_i\big) \qquad [\alpha = 1, 2] \tag{13.61}$$

are a linearly independent set and the quantities

$$S_A = \mathrm{Tr}([M]^A) = \sum_{B=1}^{3}[m_B]^A \equiv \sum_{B=1}^{3} m_{AB} \tag{13.62}$$

are two independent $SU(2)$ invariants ($S(1)$ is identically zero). At once it is clear that the general vector, which can be constructed from the M_i, has the form

$$\zeta_i = F_\alpha M_{\alpha i} \tag{13.63}$$

where the F_α are the functions of the two independent $SU(2)$ invariants. This freedom has been discussed at length by S. Gasiorowicz and D. Geffen [7].

From the point of view of field theory, it corresponds to freedom of choice of interpolating fields. Provided that $F_1(0)$ is taken to be unity and parity is respected, then all η_i so defined are equally good interpolating fields. From a geometric viewpoint, the η_i may be regarded as coordinates of points of the three-dimensional coset space manifold formed by the quotient of $SU(2) \times SU(2)$ by the vector $SU(2)$ subgroup. The freedom is then viewed as the ability to change the coordinates within a local patch near the origin.

Next we establish the transformation laws of the Killing vectors. It is sufficient to work to the lowest order in the group parameters, and we denote the transformations by

$$g : M_i \to M_i + \xi_j K^V_{ji} + \zeta_j K^A_{ji} \tag{13.64}$$

where K^V_{ji} and K^A_{ji} are Killing field components constructed from the M_i themselves. The general theory is well described by S. Coleman, J. Wess, and B. Zumino [14] and C.G. Callan, S. Coleman, J. Wess, and B. Zumino [15] and we follow the general line of their arguments in our own notation. Of course, the action under an element of the $U(1)$ subgroup is linear so that K^V_{ji} is already known, but we shall let this emerge from our calculations. It will prove convenient also to work with the combinations

$$K^L_{ij} = \frac{1}{2}[K^V_{ij} + K^A_{ij}] \tag{13.65}$$

$$K^R_{ij} = \frac{1}{2}[K^V_{ij} - K^A_{ij}] \tag{13.66}$$

corresponding to actions of the left and right chiral subgroups, respectively. The particularly important property is that these field components viewed as matrices have inverses so that

$$K^{-1}_{Lpi} K_{Liq} = \delta_{pq} = K^{-1}_{Rpi} K_{Riq} \tag{13.67}$$

$$K_{Lpi} K^{-1}_{Lqj} = \delta_{ij} = K_{Riq} K^{-1}_{Rqj} \tag{13.68}$$

follows at once from the free transitive nature of a translation action. Notice also that K_A has an inverse, but K_V is singular.

All the information required about the Killing vectors is, of course, contained in the action

$$g \exp(\xi_i A_i) = \exp(\xi'_i A_i) \exp(\eta_i V_i) \tag{13.69}$$

where, if we parametrize an arbitrary point on the manifold in the form

$$\exp(\xi_i, A_i) \equiv p(M) \tag{13.70}$$

then we can write

$$g \exp(\xi_i, A_i) = p(M') \exp(\eta_i V_i) \tag{13.71}$$

where M_i' and η_i depend upon both M_i and g and where g is an arbitrary element of the full group.

Next we apply the automorphism (induced by the parity) to Equation (13.71). If we denote by $S(g)$ the transform of an arbitrary element g, then we obtain

$$S(g)p^{-1}(M) = p(M')\exp(\eta_i V_i) \tag{13.72}$$

since the axial generators change sign. Combining this with Equation (13.71) produces

$$p^2(M') = gp^2(M)S(g^{-1}) \tag{13.73}$$

a result that has been emphasized by S. Coleman, J. Wess, and B. Zumino [14] and C.J. Isham [17]. This result has the advantage that η_i has been eliminated so that it will immediately yield information on the Killing fields alone. In the quark representation we define

$$D[p^{-1}(M) = U^{-1}(M)P^L + U(M)P^R] \tag{13.74}$$

where U is unitary and unimodular and learn that

$$\begin{aligned} U^2(M') &= \exp\left[-\frac{i}{2}\theta_i^L\sigma_i\right]U^2(M)\exp\left[-\frac{i}{2}\theta_j^R\sigma_j\right] \\ &= U^2(M) - \frac{i}{2}\theta_i^L\sigma_i U^2(M) + \frac{i}{2}\theta_i^R U^2(M)\sigma_i + \cdots \end{aligned} \tag{13.75}$$

when Equation (13.73) is applied. Now we may combine this with the form of Equation (13.64) to learn

$$-i\sigma_i U^2 = 2U^2_{,q}K^L_{ii} \tag{13.76}$$

$$iU^2\sigma_i = 2U^2_{,q}K_i^{Rq} \tag{13.77}$$

where we adopt the standard notation

$$Z_q = \frac{\partial Z}{\partial M^q} \tag{13.78}$$

for any Z and hence also find

$$K^{-1i}_{Lp} = i\,\mathrm{Tr}\left(U^2_{,p}U^{-2}\sigma^i\right) \tag{13.79}$$

$$K^{-1i}_{Rp} = -i\,\mathrm{Tr}\left(U^{-2}_{,p}U^2_{,p}\sigma^i\right) \tag{13.80}$$

since we know that these inverses exist. If we define the orthogonal transformation on the adjoint transformation M_j by

$$M_i \to R_{ij}M_j = \frac{1}{2}\,\mathrm{Tr}\left[U^{-1}\sigma_i U\sigma_j\right]M_j \tag{13.81}$$

then eliminating the derivative terms, we obtain

$$K^{L}_{iq} K^{-1}_{Rqj} = (-)\frac{1}{2}\,\mathrm{Tr}\left[U^{-2}\sigma_i U^2 \sigma_j\right] = (-)R_{ik}R_{kj}, \tag{13.82}$$

where the final step follows from the completeness relation. This is the main result of this stage of our calculations. The second step consists of expanding the whole of Equation (13.71) in the quark representation just as we treated that section of it contained in Equation (13.73).

To present the results in a tractable form we introduce

$$k^{-1}_{L\,pi} = i\,\mathrm{Tr}\left[U_{,p}U^{-1}\sigma_i\right] \tag{13.83}$$

$$k^{-1}_{R\,pi} = (-)i\,\mathrm{Tr}\left[U^{-1}U_{,p}\sigma_i\right], \tag{13.84}$$

which, as we shall see, prove to be the most important. Comparison with Equations (13.83) and (13.84) shows that the k_{ij} are components of the Killing fields for an alternative scheme in which U is replaced by its unitary unimodular square root. At once, therefore, we have a counterpart to the basic result of Equation (13.82), and we write

$$K_L + R^2 = 0 \tag{13.85}$$

$$k_L + Rk_R = 0, \tag{13.86}$$

with an obvious matrix notation. The expansion of Equation (13.71) in the quark representation then yields the results

$$2K_L k_R^{-1} = 1 + v - 2R \tag{13.87}$$

$$2K_R k_R^{-1} = 1 - v \tag{13.88}$$

$$2K_L k_L^{-1} = 1 + v \tag{13.89}$$

$$2K_R k_L^{-1} = 1 - v - 2R^{-1} \tag{13.90}$$

by similar computation to that which led to Equation (13.82) and the identity

$$U_{,q}U^{-1} + UU^{-1}_{,q} = 0 \tag{13.91}$$

is the only extra element used. Note that

$$v = K_L k_L^{-1} - K_R k_R^{-1} \tag{13.92}$$

emerges by algebra.

Simple substitution now yields at once

$$v = k^V(k^L - k^R)^{-1} = k^V(k^A)^{-1} \tag{13.93}$$

$$R = (1+v)(1-v)^{-1} \tag{13.94}$$

$$K_V = k_L + k_R = k_V \tag{13.95}$$

$$2K_A = k_A + k_V(k_A)^{-1}k_V, \tag{13.96}$$

which are the required relations. Equation (13.95) is a trivial result because we have linear transformations induced by the vector $SU(2)$ subgroup, but (as remarked previously) it is satisfying to see it arise as part of the general analysis. The remainder of these results reveal that a knowledge of k_A in a tractable form represents a complete solution for the entire nonlinear scheme we are studying.

With the decomposition given in Equation (13.55), the action of a general element g of the full group may be written as

$$\begin{aligned} g\exp(-i\xi_i A_i) &= \exp(-i\xi_i' A_i)\exp(-i\eta_i V_i) \\ &\equiv L(M')\exp(-i\eta_i V_i) \end{aligned} \tag{13.97}$$

where M_i' and η_i both depend on M_i and g. Then the primary result of the general theory is that

$$g : M_i \to M_i' \tag{13.98}$$

gives a nonlinear realization of the algebra, which is linear on the $SU(n)$ vector subgroup. Moreover if h is an element of a vector subgroup and

$$h : \Psi_\Omega \to D(h)_{\Omega\Gamma}\Psi_\Gamma \tag{13.99}$$

is a linear (unitary) representation of that subgroup, then

$$g : \Psi_\Omega \to D[\exp(-i\eta_i V_i)]_{\Omega\Gamma}\Psi_\Gamma \tag{13.100}$$

gives a realization of the full group. Notice that this latter transformation is linear in Ψ but nonlinear (through η_i) in the M_i when g is not the vector subgroup. Fields that transform according to Equation 13.101 are called standard fields and it is important to understand that by a suitable redefinition of coordinates any nonlinear realisation of $SU(2) \times SU(2)$, which is linear on the vector subgroup, can be brought into this standard form. In practice the most useful result is that, if one has a linear irreducible (unitary) representation of $SU(2) \times SU(2)$ such that

$$g : N_\Omega \to D[g]_{\Omega\Gamma}N_\Gamma \tag{13.101}$$

$$\text{then} \quad \Psi_\Omega(M) = D[L^{-1}(M)]_{\Omega\Gamma}N_\Gamma \tag{13.102}$$

transform as the components of standard fields.

It is now clear that there are three classes of fields to consider:

1. Linear representations, which may be built up in the usual way as multispinors with the transformation laws defined by Equation (13.56). These will not be treated in more detail.
2. Vectors M_i transforming as the adjoint representation of $SU(2)$ with a nonlinear transformation law under chiral action specified by Equation (13.97). These will allow a description of the massless Goldstone bosons (pions, etc.) corresponding to the axial degrees of freedom spontaneously violated. The specification of invariants constructed (nonlinearly) from these is most important and will be exhibited later.
3. Standard fields, which appear linearly in their transformation laws, but with nonlinear functions of the M_i induced according to Equations (13.101) and (13.98). These are important in describing matter (e.g., nucleons) interacting with the Goldstone bosons as chiral matter. Once more, the specification of the corresponding invariants is most important and will be given later.

The technical problem of finding the invariants is solved in Barnes, Dondi, and Sarkar [6]. A crucial step is the resolution of the powers of the matrix M in the form

$$[M]^A = [m_B]^A P_B \equiv M_{AB} P_B \tag{13.103}$$

where the P_B are two Hermitian matrices, each 2×2, with the properties

$$P_A P_B = \delta_{AB} P_B \quad (\text{ no sum}) \tag{13.104}$$

$$\text{Tr}(P_A) = 1 \tag{13.105}$$

and

$$\sum_{A=1}^{2} P_A = \mathbf{1} \tag{13.106}$$

where this $\mathbf{1}$ is the unit (2×2) matrix. Although the P_A are not in general diagonal, the projection operator

$$P_{Ai} = \frac{1}{2}\,\text{Tr}(P_A \lambda_i) \tag{13.107}$$

and

$$(P_A)_{MN} = P_{Ai}(\lambda_i)_{MN} + \frac{1}{n}\delta_{MN} \tag{13.108}$$

where, because the P_A are complete, it follows that

$$\sum_{A=1}^{2} P_{Ai} = 0 \tag{13.109}$$

and introducing

$$P_{Ai} = \sqrt{2}\left[P_{Ai} - (1+\sqrt{2})^{-1}\right]P_{2i} \tag{13.110}$$

with

$$\sqrt{2}P_{Ai} = p_{Ai} + \frac{1}{\sqrt{2}}p_{2i} \tag{13.111}$$

establishes that $p_{\mu i}$ are orthonormal.

The second-rank tensors defined by the M_i are conveniently handled by an extension of these ideas and fall into two classes. One such class is formed by the two independent tensors defined by

$$(P_{AB}) \equiv P_{AiBj} \equiv \frac{1}{2}\,\mathrm{Tr}(P_A\lambda_i P_B\lambda_j) \qquad (A \neq B) \tag{13.112}$$

and

$$I_{ij} = \frac{1}{2}\,\mathrm{Tr}(P_A\lambda_i P_B\lambda_j), \tag{13.113}$$

which have the properties

$$II = I \tag{13.114}$$

$$IP_{AB} = 0 = P_{AB}I \tag{13.115}$$

$$P_{AB}P_{CD} = \delta_{AC}\delta_{BD}P_{AB} \qquad (\text{ no sum}) \tag{13.116}$$

in terms of the matrix notation of the last section. Moreover, these are all Hermitian matrices and the trace of each P_{AB} is unity. Since it is easy to show also that

$$\sum_{A\neq B}{}' P_{AB} = \mathbf{1} - I \tag{13.117}$$

where the sum is over all A and B but excluding terms with $A = B$, this gives a projection operator resolution in one sector of the space of these second-rank tensors and so I will decompose further. The second class of tensors may be identified with the independent matrices with components

$$(p_{\alpha\beta})_{ij} \equiv p_{\alpha i}p_{\beta j}, \tag{13.118}$$

which span the subspace of 3×3 matrices projected out on multiplication by I from both sides and which are therefore orthogonal to the subspace in which the P_{AB} lie. Since the $p_{\alpha i}$ are orthonormal, the multiplication law for the $p_{\alpha\beta}$ is

$$p_{\alpha i}p_{\beta j} = \delta_{\beta\gamma}p_{\alpha\delta}. \tag{13.119}$$

It has been established by Barnes and Delbourgo [16] that all the independent second-rank tensors which can be constructed from the M_i are spanned by the three independent $p_{\alpha\beta}$ and P_{AB}.

The most general unitary unimodular matrix U constructed from the M_i may be written in the form

$$U = U_A P_A = \exp\left[-\frac{i}{2}\theta_A\right] \quad \text{where} \quad \sum_{A=1}^{n} \theta_A = 0 \tag{13.120}$$

but the θ_A are otherwise completely arbitrarily independent functions of the independent $SU(2)$ invariant S_A subject to the considerations of parity and weak field limits as mentioned before. This effective arbitrary function of the invariant is characteristic of the general solution and will persist throughout this work.

It has been conventional to define

$$\sqrt{2}\phi_A = m_A - \frac{m/2}{\sqrt{2}} \tag{13.121}$$

with

$$m_A = \sqrt{2}(\phi_A + \phi) \tag{13.122}$$

so that, extending the notation used previously

$$M_i = \phi p_i \tag{13.123}$$

$$\phi_i = p_i \tag{13.124}$$

follow immediately. Similarly, defining

$$\sqrt{2}\psi_A = \theta_A - \theta \tag{13.125}$$

$$\theta_A = \sqrt{2}\psi_A + \theta \tag{13.126}$$

the ψ_A may be treated as an independent (arbitrary) function of the ϕ, which then serves as the independent variant.

The transformation laws for all realizations are now given by Dondi and Sarkar [6] in closed form and in terms of simple functions. Restricting attention to first-order derivatives of the field with respect to space and time, and also restricting attention to a study of the Goldstone boson fields M_i and the standard fields, the results can be given in terms of the general analysis of Barnes and Delbourgo [16] and Isham [17]. There are two important results. First, although $\partial_\mu M_i$ and $\partial_\mu \Psi_\Gamma$ do not transform as standard fields, the covariant derivatives

$$D_\mu M_i = a_{\mu i} \tag{13.127}$$

$$D_\mu \Psi_\Gamma = \partial_\mu \Psi_\Gamma - i v_{\mu i} (T_i)_{\Gamma\Omega} \Psi_\Omega \tag{13.128}$$

where under $SU(2)$

$$\Psi \rightarrow \Psi_\Gamma - i\theta_i (T_i)_{\Gamma\Omega} \Psi_\Omega \tag{13.129}$$

and where

$$L^{-1}(M)\partial_\mu L(M) = \exp(-\xi_i A_i)\partial_\mu \exp(-\xi_i A_i) = v^i_\mu V_i + a^i_\mu A_i \tag{13.130}$$

have precisely this property. Second, they show that the most general Lagrangian of the type under consideration may be written as a function of the standard fields Ψ, $D_\mu \Psi$, and $D_\mu M_i$ only; that is, the M_i will not appear explicitly and the Goldstone bosons will be massless. It then follows that the Lagrangian so formed will be invariant under $SU(2) \times SU(2)$ if and only if it is constructed to be invariant under the $SU(2)$ vector subgroup. This latter requirement is achieved by index saturation.

The result given in Barnes, Dondi, and Sarkar [6] (now dropping the chiral projectors and normalizing for this problem) takes the concrete form

$$D_\mu M_i = \left\{ \frac{\partial \psi_\beta}{\partial \phi_\gamma}(P_{\gamma\beta})_{ik} + \sum_{A \neq B}{}' \frac{\sqrt{2}}{\phi_A - \phi_B} \sin \left[\frac{\psi_A - \psi_B}{\sqrt{2}}\right] (P_{AB} + P_{BA})_{ik} \right\} (\partial_\mu M_k) \tag{13.131}$$

and represents a complete specification of the required Lagrangian in simple closed form. This can be rewritten in the form

$$D_\mu M_i = \left[\frac{d\theta}{d\phi} n_k n_i + \frac{\sin\theta}{\phi}(\delta_{ki} - n_k n_i) \right], \tag{13.132}$$

which is perhaps a simpler form, already seen by the reader.

Noting the appearance of the projection operators, and realizing that the invariant Lagrangian density L is found by index saturation, it is straightforward to see that it is proportional to a constant

$$k = \left(\frac{d\theta}{d\phi}\right)^2 n_A n_B + \frac{\sin^2\theta}{\phi^2}(\delta_{AB} - n_A n_B) \tag{13.133}$$

where we have now taken the indices to the A range, which would have been confusing in the previous equation. The condition for this to be Hermitian is obviously

$$\left(\frac{d\theta}{d\phi}\right)^2 = \frac{\sin^2\theta}{\phi^2} \tag{13.134}$$

with the solution

$$\phi = c \tan\frac{\theta}{2} \tag{13.135}$$

being the one conventionally chosen, where c is a constant. When $c = 1$, the invariant Lagrangian density is

$$\mathcal{L} = \frac{(\partial_\mu \xi)(\partial^\mu \overline{\xi})}{2[r^2 + \xi\overline{\xi}]} \tag{13.136}$$

where $\xi = r(M_1 + iM_2)$ is used to emphasize the constant radius of the 2-sphere.

Using the geometric formulation of Isham [17] gives the coset space metric in the form related to the covariant derivatives as

$$g_{ij}(\partial_\mu M_i)(\partial^\mu M_j) = (D_\mu M_i)(D^\mu M_i) \tag{13.137}$$

and we have normalized g_{ij} to δ_{ij} in the limit of zero fields. In matrix notation this yields

$$g = \frac{1}{4}\left[p_{\beta\lambda}\frac{\partial\psi_\alpha\partial\psi_\alpha}{\partial\phi_\beta\partial\phi_\lambda}\right] + \frac{1}{4}\left[\sum_{A\neq B}{}' \frac{2}{(\phi_A - \phi_B)^2}(P_{AB} + P_{BA})\sin^2\left[\frac{\psi_A - \psi_B}{\sqrt{2}}\right]\right] \tag{13.138}$$

immediately because of the orthonormality.

This is, perhaps, an ideal moment to tell first-year research students about what may actually be possible. Soon after completing his first-year courses and examinations, Chris Isham came to ask for suggestions for a research project. Professor Paul Matthews and I were not quite prepared for this! So Paul explained to him what was possibly the problem of the decade and which neither of us had any idea of how to solve, and off Chris went to try his hand. A few days later Chris caught me alone and asked two technical questions, which I was able to answer. Soon afterward he came into the offce where Paul and I were working, put down a pile of paper and said, "Is this what you are looking for?" A glance showed us, although we did not understand the new nonlinear mathematics that he used, that he had solved the problem. Paul said, "Wonderful, now go away and put in the fermions." When Chris left, Paul wrote a title page, with the author "C.J. Isham" and handed it to the secretary to type for publication. Not long afterward Chris came with the next paper. Chris and I started writing a series of related papers. One day he showed me a letter and asked if I understood it. It was from one of the great East Coast universities in the United States, offering him a tenured professorship at a huge salary, and the right to appoint four new members of academic staff without consulting anyone. I advised him to show the letter to Paul and Abdus Salam. Chris wrote his PhD thesis and passed the interview, still before the end of his first year of research. After a year working at the Trieste Institute, he returned to Imperial College as a lecturer. He is still there as a professor of theoretical physics, although he is now much better known for his research on general relativity. We became very good friends and are still in contact.

But enough of this; let us turn our attention to field redefinitions. We have alluded to the generality and utility of our parametrization. Now it is time to see the scheme in action. Consider first the coordinates resulting from a constraint on the singlet in the chiral $SU(2)$ triplet representation. In chiral

$SU(2)$, one introduces a multiplet

$$\Phi = \pi_A \sigma_A \gamma_5 + \sigma \tag{13.139}$$

transforming as a $q\overline{q}$ bispinor, where

$$q \to q + \frac{\theta}{2}\sigma_3 q - i\frac{\phi_A}{2}\tau_A(i\gamma_5)q \tag{13.140}$$

as we have seen previously. It is trivial to see that the group action on this three-dimensional multiplet is

$$\pi_A \to \pi_A + \varepsilon_{A3B}\theta\sigma_B + \phi_A\sigma \tag{13.141}$$

$$\sigma \to \sigma - \phi_A\pi_A \tag{13.142}$$

and we recognize a normal linear representation in which the scalar field is a $U(1)$ singlet. In this scheme the nonlinearity results from imposing the chiral $SU(2)$ invariant constraint

$$\sigma^2 + \pi_A\pi_A = f_\pi^2 \tag{13.143}$$

where f_π^2 is constant, to eliminate the σ field. The transformation law is then

$$\pi_A \to \pi_A + \varepsilon_{A3B}\theta\pi_\pi + \phi_B\left[f_\pi^2 - \pi^2\right]^{\frac{1}{2}}\delta_{AB} \tag{13.144}$$

where $\pi^2 = \pi_A\pi_A$ and we have arbitrarily selected the positive square root.

We emphasize that this is an example of the transformation laws derived earlier for a particular choice of our arbitrary invariant function $\theta(\pi)$. The simple form of this transformation law, together with the intuitive feeling for the nonlinearity arising from the constraint, have made this a popular choice in the literature. However, we now return to the stereographic choice of coordinates used by Zumino [4] to allow the introduction of supersymmetry by emphasizing the Kähler properties of the 2-sphere. Things now look very different. The two real coordinates on the sphere (pseudoscalar fields) are replaced by a single complex variable z. Now the metric is an Hermitian form and the nonzero components are written as

$$g_{z\overline{z}} = \frac{\partial^2 V}{\partial z \partial \overline{z}} \tag{13.145}$$

where V is a potential function so that the usual cross-derivate constraints [6] are satisfied for this to be a Kähler manifold. In this framework, the nonlinear transformation law for the coordinates takes the form

$$z \to z + i\theta z + \frac{c\omega}{2} + \frac{\overline{\omega}z^2}{2c} \tag{13.146}$$

$$\text{where} \quad \omega = \phi_1 + i\phi_2. \tag{13.147}$$

c is a constant (identifiable as $2f_\pi$) and we note that this transformation law is holomorphic in z. Of course, it is simple to change to a more familiar pair of real variables, now $x_A(A = 1, 2)$, by setting

$$z = x_1 + i x_2 \tag{13.148}$$

and we find that Equation (13.142) yields

$$x_A \to x_A + \varepsilon_{A3B}\theta x_B + \phi_B \left[\frac{\delta_{AB}(c^2 - x^2)}{2c} + \frac{x_A x_B}{c} \right] \tag{13.149}$$

$$\text{where} \quad x^2 = x_A x_A \tag{13.150}$$

and Equation (13.147) has been used. By comparing the δ_{AB} terms in Equations (13.64) and (13.142) we discover

$$\pi \cot(\theta) = \left[f_\pi^2 - \pi^2\right]^{\frac{1}{2}} \tag{13.151}$$

and similarly comparing Equations (13.144) and (13.151)

$$2cx \cot(\theta) = c^2 - x^2 \tag{13.152}$$

emerges. Direct comparison of these last two results gives

$$\pi^2 = \frac{4c^2 f_\pi^2 x^2}{[c^2 + x^2]^2} \tag{13.153}$$

or equivalently

$$\pi_A = \frac{2c f_\pi x_A}{c^2 + x^2} \tag{13.154}$$

as the connection between the two coordinate systems. We note again that c may be identified with $2f_\pi$ when the equality of the two coordinate systems is transparent in the small field limit. This simple example, I hope, makes clear the advantage of working with the general coordinate treatment.

If you have no background or interest in particle physics, move on to another chapter. I am not sure whether this section should be included at all, but it is short, so here we go.

Think about the chiral sphere $\frac{SU(2)}{U(1)}$ as previously described. In the particle physics interpretation there are two pseudoscalar Goldstone bosons with equal and opposite unit charges $\pm e$. They must be massless. Of course, in the real world, experiment tells us that they have masses that are not zero, but are the lightest of all masses in the hadronic sector. Now, recall the Borel theorem of Chapter 10, which tells us that because the centralizer of the $U(1)$ in $SU(2)$ is just itself we have a single supersymmetry. In practice, this integer controls the number of supersymmetries that can be imposed. This becomes exceptionally important in advanced topics, such as the Maldacena conjecture. It allows what might be thought of as improvements to this already very important topic.

Just in case even a single reader has been inspired to work in theoretical high energy particle physics, I am adding a few extra remarks. To other readers my apologies, but you can read this section too if you feel so inclined.

The first thing, of course, is to find a good supervisor (interested in these topics) at a good university physics or mathematics department and then to get yourself a support grant from the appropriate research council. It is much easier to survive on these grants than it was just a few years ago.

Then despite having years of training to do, invest in a copy of the book *Harmonic Superspace* by A.S. Galperin, E.A. Ivanov, V.I. Ogievetsky, and E.S. Sokatchev [18], who worked together at the joint Institute for Nuclear Research in Dubna, Russia, where harmonic superspace was born. The first draft of the book was written some years ago when all four authors were working together at the Bogoliubov Laboratory of Theoretical Physics. I find in every generation there is some outstanding worker whose research papers must be followed at almost any cost. Ogievetsky was one of my favorites, and his death was a huge loss. I am confident that his colleagues would not mind my making special mention of him.

In this book you can learn how our simple sphere can be used to extend the index of Kähler manifolds and therefore the number of supersymmetries. You can also learn how to improve upon the famous Maldacena Ads/CFT conjecture [19]. In fact, the whole book is a real treasure-house in the subject.

References

1. P. Chang and F. Gürsey, *Phys. Rev.* 164 (1967): 1752.
2. S. Weinberg, *Phys. Rev. Lett.* 18 (1967): 188.
3. S. Weinberg, *Phys. Rev. Lett.* 166 (1968): 1568.
4. B. Zumino, *Phys. Lett.* 87B (1979): 203.
5. K.J. Barnes, J. Generowicz, and P. Grimshaw, *J. Phys. A: Math. Gen.* 29 (1996): 4457.
6. K.J. Barnes, P. Dondi, and S. Sarkar, *Proc. R. Soc.* A330 (1972): 389.
7. S. Gasiorowicz and D. Geffen, *Rev. Mod. Phys.* 41 (1969): 531.
8. S. Weinberg, in *Proceedings of the 14th International Conference on High Energy Physics*, edited by J. Prentki and J. Steinberger, CERN, Geneva, 1968, p. 253.
9. J. Gasser and H. Leutwyler, *Ann. Phys. (NY)* 158 (1984): 142.
10. U.-G. Meissner, *Rep. Prog. Phys.* 56 (1993): 903.
11. T. Clark and W.T. Veldhuis, *Nucl. Phys.* B426, no. (2) (1994): 385.
12. K.J. Barnes, D. Ross, and R. Simmons, *Phys. Lett.* B338 (1994): 457.
13. L. Michel and L.A. Radicati, *Symmetry Principles at High Energy: 5th Coral Gables Conference*. W.A. Benjamin, New York, 1968, p. 9.
14. S. Coleman, J. Wess, and B. Zumino, *Phys. Rev.* 177 (1969): 2239.
15. G.G. Callan, S. Coleman, J. Wess, and B. Zumino, *Phys. Rev.* 177 (1969): 2247.
16. K.J. Barnes and R. Delbourgo, *J. Phys. A: Gen. Phys.* 5 (1972): 1043.
17. C.J. Isham, *Nuovo Cimento* 59A (1969): 356.

18. A. Galperin, E. Ivanov, V. Ogievetsky, and E. Sokatchev, *Harmonic Superspace*, Mongraphs on Mathematical Physics, Cambridge University Press, Cambridge, 2001.
19. J.M. Maldacena, *Adv. Theor. Math. Phys.* 2 (1998): 231.

Problems

13.1 Read through Equations (13.1) to (13.4). Now close your notes and find the transformation law for π_A.

13.2 Check that you can recover the nonlinear transformation law for z in Equation (13.6).

13.3 Read through Equations (13.8) to (13.14). Now close your notes and find the expression for π_A connecting the coordinate systems.

13.4 Without using your notes, explain why in the embedding of $SU(2)$ into $SU(2) \times SU(2)$ it is crucial to express the parametrization of the quotion coset space in the form

$$\hat{L} = \exp\left\{-\frac{1}{2} i\xi n_i \sigma^i (i\gamma_5)\right\}$$

and say what each of the quantities in this expression is and what role it plays.

13.5 Show how the projection operators $P_L = \frac{1}{2}(1 + i\gamma_5)$ and $P_R = \frac{1}{2}$ $(1 - i\gamma_5)$ work.

13.6 Read through your notes from Equations (13.1) to (13.12). Now close your notes and write how the form of $g = \exp(-i\xi_i A_i)\exp(-i\zeta_i V_i)$ is reached.

13.7 Read through your notes from Equations (13.21) to (13.25). Now close your notes and work through this for yourself.

13.8 Read through your notes from Equations (13.26) to (13.39). Now close your notes and try to do this for yourself. This is a hard section. You may refer to your notes whenever you need, but make sure that you can do all the steps eventually.

13.9 Read through your notes for Equations (13.40) to (13.53). Now close your notes and reproduce this material for yourself.

13.10 Read through your notes from Equations (13.1) to (13.5). Now close your notes and do this for yourself.

13.11 Read through your notes from Equation (13.6) until you have seen the three classes of fields we need to consider. Now close your notes and do this for yourself.

13.12 The next section from Equations (13.8) to (13.24) is generally thought to be difficult. Please read it until you feel sure that you understand it.

13.13 Read your notes from Equation (13.25) to Equation (13.31) and make sure that you understand it.

13.14 Read your notes from Equation (13.32) to Equation (13.35) and the next paragraph. Make sure you understand it and why it is being done.

13.15 Using the results of Problem (13.14) and inventing your own normalization, derive the expression for $D_\mu M_i$ in Equation (13.128).

13.16 Read through your notes from Equation (13.37) to Equation (13.41). Make sure you understand why this last equation represents the surface of the two-dimensional sphere.

13.17 Read the section of your notes from Equation (13.1) until you know what happened to Professor Chris Isham. Think about this.

13.18 Read your notes from Equation (13.44) to Equation (13.49) until you are sure you can do this for yourself.

13.19 Make sure that you understand the stereographic coordinates used by Zumino.

13.20 Check Equation (13.147) for yourself. Check the law is holomorphic in z.

13.21 Work out the expression for x_A in Equation (13.42) for yourself.

14

Beyond the Standard Model

The world we see is Poincaré invariant. It also has *internal symmetries*—both *global* and *local*—the latter giving rise to interactions (*forces*) described by gauge theories. Can there be more? Can we mix Poincaré and internal? Only by supersymmetry, which mixes bosons with fermions.

We need to establish notation. The basic objects we need will be *spinor representations* of the $SO(1,3)$ Lorentz group. We take the metric $g^{\mu\nu}$ defined by

$$\begin{aligned} g^{00} &= 1 \\ g^{ij} &= -\delta^{ij} \\ g^{0i} &= 0. \end{aligned} \tag{14.1}$$

The primary tool we need is the Clifford algebra. We introduce the anticommutator

$$\{\gamma^{\mu}, \gamma^{\nu}\} = 2g^{\mu\nu} \tag{14.2}$$

where as we know the Dirac matrices have a unique representation (up to equivalence) as 4×4 matrices. By alternately forming commutators and anticommutators we find all 16 independent matrices. In this notation γ^0 is Hermitian and the γ^i are antihermitian.

One representation, with useful low energy properties, is given in terms of the Pauli matrices

$$1 = \begin{pmatrix} 1 & 0 \\ 0 & 1 \end{pmatrix} \sigma_1 = \begin{pmatrix} 0 & 1 \\ 1 & 0 \end{pmatrix} \sigma_2 = \begin{pmatrix} 0 & -i \\ i & 0 \end{pmatrix} \sigma_3 = \begin{pmatrix} 1 & 0 \\ 0 & -1 \end{pmatrix}, \tag{14.3}$$

which have products

$$\sigma^i \sigma^j = \delta^{ij} + i\varepsilon^{ijk}\sigma^k\% \tag{14.4}$$

in terms of the Kronecker delta and Levi-Civita tensor. We can take

$$\gamma^0 = \begin{pmatrix} 1 & 0 \\ 0 & -1 \end{pmatrix} \gamma^i = \begin{pmatrix} 0 & \sigma^i \\ -\sigma^i & 0 \end{pmatrix}, \tag{14.5}$$

DOI: 10.1201/9781439895207-14

which have the required properties. Building the rest of the Dirac matrices as suggested gives sequentially

$$\sigma^{\mu\nu} = \frac{i}{2}[\gamma^\mu, \gamma^\nu], \tag{14.6}$$

which in the above representation take the form

$$\sigma^{0i} = \begin{pmatrix} 0 & i\sigma^i \\ i\sigma^i & 0 \end{pmatrix} \sigma^{ij} = \varepsilon^{ijk} \begin{pmatrix} \sigma^k & 0 \\ 0 & \sigma^k \end{pmatrix}. \tag{14.7}$$

Then, realizing that the maximum number of Dirac matrices in a product is four, we introduce

$$\begin{aligned} \gamma^5 &= \gamma^0\gamma^1\gamma^2\gamma^3 \\ &= \frac{1}{4!}\varepsilon_{\mu\nu\rho\lambda}\gamma^\mu\gamma^\nu\gamma^\rho\gamma^\lambda. \end{aligned} \tag{14.8}$$

Here the notation is $\varepsilon_{0123} = 1$.

In this representation we find

$$\gamma^5 = \begin{pmatrix} 0 & -i \\ -i & 0 \end{pmatrix}, \tag{14.9}$$

which is antihermitian, squares to -1, and has

$$\{\gamma^5, \gamma^\mu\} = 0 \tag{14.10}$$

as a property, which clearly "extends" the Clifford algebra.

We then define

$$\begin{aligned} \sigma^{\mu 5} &= i[\gamma^\mu, \gamma^5] \\ &= \frac{i}{3!}\varepsilon^{\mu\nu\rho\lambda}\gamma_\nu\gamma_\rho\gamma_\lambda \end{aligned} \tag{14.11}$$

to complete our set of 16 basic 4×4 matrices. In our representation these take the forms

$$\begin{aligned} \sigma^{05} &= \begin{pmatrix} 0 & 1 \\ -1 & 0 \end{pmatrix} \\ \sigma^{i5} &= \begin{pmatrix} \sigma^i & 0 \\ 0 & -\sigma^i \end{pmatrix}, \end{aligned} \tag{14.12}$$

where we can easily check that the notation is designed so that we have

$$\gamma^0\Gamma_R^\dagger\gamma^0 = \Gamma_R \tag{14.13}$$

for $R = 1, \ldots 16$. See Chapter 4 for the γ matrix product table.

At this stage we had better think about the Poincaré group to put our notation into a physical context. The transformations consist of translations in space–time, together with *spatial rotations* and *Lorentz boosts* to different velocities. We can write this in the form

$$x'^{\mu} = \Lambda^{\mu}_{\ \nu} x^{\nu} + a^{\mu} \tag{14.14}$$

where the a^{μ} and $\Lambda^{\mu}_{\ \nu}$ are constants, and the constraint equation

$$g_{\mu\nu} = g_{\rho\lambda} \Lambda^{\rho}_{\ \mu} \Lambda^{\lambda}_{\ \nu} \tag{14.15}$$

ensures invariance of our metric.

As Lie assures us, we can work without loss of generality in infinitesimal form

$$\begin{aligned} \Lambda^{\mu}_{\ \nu} &= \delta^{\mu}_{\ \nu} + \omega^{\mu}_{\ \nu} \\ a^{\mu} &= \varepsilon^{\mu} \end{aligned} \tag{14.16}$$

with constant small parameters ε^{μ} and $\omega^{\mu\nu}$, and the constraint equation simply tells us that

$$\omega^{\mu\nu} = -\omega^{\nu\mu} \tag{14.17}$$

so we have four parameters ε^{μ} for the translations and six parameters $\omega^{\mu\nu}$, where ω^{ij} describe rotations and ω^{0i} describe boosts. We can view a small group element as

$$g = 1 - \frac{i}{2}\omega_{\alpha\beta} M^{\alpha\beta} + i\varepsilon_{\alpha} P^{\alpha} \tag{14.18}$$

and expanding our transformation as

$$\begin{aligned} x'^{\mu} &= x^{\mu} + \omega^{\mu}_{\ \nu} x^{\nu} + \varepsilon^{\mu} \\ &= x^{\mu} + \omega_{\alpha\beta} g^{\mu\alpha} x^{\beta} + \varepsilon_{\alpha} g^{\mu\alpha} \\ &= x^{\mu} + \frac{1}{2}\omega_{\alpha\beta}[g^{\mu\alpha} x^{\beta} - g^{\mu\beta} x^{\alpha}] + \varepsilon_{\alpha} g^{\mu\alpha} \end{aligned} \tag{14.19}$$

we discover that the generators satisfy the algebra of the Poincaré group by having the action on fields (not coordinates!) specified by

$$P^{\alpha} = i\partial^{\alpha} \tag{14.20}$$

$$\text{and} \quad M^{\alpha\beta} = i(x^{\alpha}\partial^{\beta} - x^{\beta}\partial^{\alpha}). \tag{14.21}$$

It is then easy to confirm the algebra as

$$\begin{aligned} [P^{\mu}, P^{\nu}] &= 0 \\ [M^{\mu\nu}, P^{\rho}] &= i(g^{\nu\rho} P^{\mu} - g^{\mu\rho} P^{\nu}) \\ [M^{\mu\nu}, M^{\rho\lambda}] &= i(M^{\mu\lambda} g^{\nu\rho} - M^{\mu\rho} g^{\nu\lambda} + M^{\nu\rho} g^{\mu\lambda} - M^{\nu\lambda} g^{\mu\rho}). \end{aligned} \tag{14.22}$$

We note that the translations commute, and form a vector representation of the homogeneous Lorentz group whose generators satisfy the algebra in the final line.

Recall how the states are found in quantum mechanics. We look for a complete commuting set of observables. An obvious first choice are the P^μ since they commute with each other. So we set

$$P^\mu|p^\mu\rangle = p^\mu|p^\mu\rangle \tag{14.23}$$

and recognize that $P^2 = P^\mu P_\mu$ is an invariant. Now we can distinguish the massive and massless cases.

Massive Case

For the $m \neq 0$ case we work in the rest frame where $E = p^0 = m$ and $\vec{p} = 0$. We then ask for members of the algebra commuting with the commuting observables we already have and discover the *little group*, also known as the *stability group*, generated by the M^{ij}. These are rotations of course, and by writing

$$M^{ij} = \varepsilon^{ijk} J^k \tag{14.24}$$

$$\text{or} \quad J^i = \frac{1}{2}\varepsilon^{ijk} M^{jk} \tag{14.25}$$

we reveal the algebra

$$[J^i, J^j] = i\varepsilon^{ijk} J^k \tag{14.26}$$

of $SU(2)$ or $SO(3)$ or angular momentum. We now add J_z and $J^2 = J^i J^i$ to the complete commuting set of observables in the form

$$\begin{aligned} J^2|j,m\rangle &= j(j+1)|j,m\rangle \\ J_z|j,m\rangle &= m|j,m\rangle \end{aligned} \tag{14.27}$$

where j is an integer or $\frac{1}{2}$ integer, and m takes all values between $-j$ and j in integer steps. We now have the massive particles we know and love.

Massless Case

In the $m = 0$ case we go to the frame where $p^\mu = (E; 0, 0, E)$ and discover that the little group is now best described by introducing

$$W^\mu = \frac{1}{2}\varepsilon^{\mu\nu\rho\lambda} P_\nu M_{\rho\lambda} \tag{14.28}$$

where we note that $P_\mu W^\mu \equiv 0 \Rightarrow W^\mu$ has only three independent components. The algebra of the little group is now (in this frame)

$$\begin{aligned} [W^1, W^2] &= 0 \\ [(W^3/E), W^2] &= iW^1 \\ [(W^3/E), W^1] &= -iW^2, \end{aligned} \tag{14.29}$$

which we recognize as a [2] Euclidean algebra, where the W^1 and W^2 are translations and $\frac{W^3}{E}$ is the rotation in the plane. This is not a compact group algebra (nor is the Lorentz group, of course) and it can only have finite dimensional representations if they are nonunitary. In practice we have only found particles in which W^1 and W^2 have zero eigenvalues. (They could in principle be continuous.) We then define the helicity, a pseudoscalar, by

$$\lambda = \frac{n_\mu W^\mu}{n_\mu P^\mu} \tag{14.30}$$

where n_μ is an arbitrary vector with $n_\mu P^\mu \neq 0$. We usually take n_μ along the time direction when

$$\lambda = \frac{\vec{p} \cdot \vec{J}}{|\vec{p}|} \tag{14.31}$$

and we speak of the component of the angular momentum, or *spin*, along the three-momentum. We now have the massless photon with two polarization states, and so on.

Projection Operators

Now let's return to our Dirac matrices, and consider the transformations of Dirac four-component spinors (either wave functions or fields—just reverse the sign of the parameter if you have a vector operator) under the Lorentz group. We write

$$\begin{aligned} \psi'_\alpha(\Lambda x) &= S_\alpha^{\ \beta} \psi_\beta(x) \\ &= \psi_\alpha(x) - \frac{i}{4}\omega_{\mu\nu}\,(\Sigma^{\mu\nu})_\alpha^\beta\,\psi_\beta(x) + \ldots \end{aligned} \tag{14.32}$$

and need the $\Sigma^{\mu\nu}$ to be a realization of the $M^{\mu\nu}$. Looking at the Dirac matrix multiplication table we realize that

$$\Sigma^{\mu\nu} \equiv \sigma^{\mu\nu} \tag{14.33}$$

for the spinors. Now we notice also that

$$[\gamma^5, \sigma^{\mu\nu}] \equiv 0 \tag{14.34}$$

so we define projection operators

$$P_L = \frac{1 + i\gamma^5}{2} \quad \text{and} \quad P_R = \frac{1 - i\gamma^5}{2} \tag{14.35}$$

with the following properties

$$\begin{aligned} P_L P_L &= P_L \\ P_R P_R &= P_R \\ P_L P_R &= P_R P_L = 0 \\ P_L + P_R &= 1. \end{aligned} \tag{14.36}$$

Weyl Spinors and Representation

Obviously these commute with the $\sigma^{\mu\nu}$, so they can be used to decompose the Dirac spinor into left and right (2 × 1) Weyl spinors,

$$\psi_L = P_L \psi \tag{14.37}$$

$$\psi_R = P_R \psi, \tag{14.38}$$

which are actually two-component spinors. We can easily see that ψ_L and ψ_R transform using, respectively,

$$\begin{aligned} \sigma_L^{\mu\nu} &= P_L \sigma^{\mu\nu} P_L \\ &= \frac{1}{2}\left(\sigma^{\mu\nu} - \frac{i}{2}\varepsilon^{\mu\nu\alpha\beta}\sigma_{\alpha\beta}\right) \end{aligned} \tag{14.39}$$

$$\begin{aligned} \text{and} \quad \sigma_R^{\mu\nu} &= P_R \sigma^{\mu\nu} P_R \\ &= \frac{1}{2}\left(\sigma^{\mu\nu} + \frac{i}{2}\varepsilon^{\mu\nu\alpha\beta}\sigma_{\alpha\beta}\right). \end{aligned} \tag{14.40}$$

There is an alternative representation of the Dirac matrices, usually called the *Weyl representation* or the *chiral representation*, which makes this easier to visualize. Here we have

$$\gamma^0 = \begin{pmatrix} 0 & -1 \\ -1 & 0 \end{pmatrix} \quad \gamma^i = \begin{pmatrix} 0 & \sigma^i \\ -\sigma^i & 0 \end{pmatrix} \quad \gamma^5 = \begin{pmatrix} -i & 0 \\ 0 & i \end{pmatrix}$$
$$\sigma^{0i} = \begin{pmatrix} i\sigma^i & 0 \\ 0 & -i\sigma^i \end{pmatrix} \quad \sigma^{ij} = \varepsilon^{ijk}\begin{pmatrix} \sigma^k & 0 \\ 0 & \sigma^k \end{pmatrix} \quad \sigma^{05} = \begin{pmatrix} 0 & 1 \\ -1 & 0 \end{pmatrix}$$
$$\sigma^{i5} = \begin{pmatrix} 0 & -\sigma^i \\ -\sigma^i & 0 \end{pmatrix}. \tag{14.41}$$

Obviously

$$P_L = \left(\frac{1+i\gamma^5}{2}\right) = \begin{pmatrix} 1 & 0 \\ 0 & 0 \end{pmatrix} \tag{14.42}$$

$$\text{and} \quad P_R = \left(\frac{1-i\gamma^5}{2}\right) = \begin{pmatrix} 0 & 0 \\ 0 & 1 \end{pmatrix} \tag{14.43}$$

$$\text{so that} \quad \psi = \begin{pmatrix} \psi_L \\ \psi_R \end{pmatrix}. \tag{14.44}$$

We can then see that

$$\psi_L \to \psi_L + \frac{i\omega^{0i}}{2}(i\sigma^i)\psi_L - \frac{i\omega^{ij}}{4}\varepsilon^{ijk}\sigma^k\psi_L$$
$$\psi_R \to \psi_R + \frac{i\omega^{0i}}{2}(-i\sigma^i)\psi_R - \frac{i\omega^{ij}}{4}\varepsilon^{ijk}\sigma^k\psi_R \tag{14.45}$$

that is, they transform the same way under rotations, but not under boosts! Notice the extra i factor with ω^{0i}—this is $SO(1,3)$ or $SL(2,C)$ not $SU(2) \times SU(2)$. The representations are finite dimensional only by being nonunitary; this is a noncompact group.

We are familiar with this through the use of the *adjoint spinor*

$$\overline{\psi}^{\alpha} = (\psi_\beta)^\dagger(\gamma^0)^{\alpha}_{\beta}, \tag{14.46}$$

which transforms as

$$\overline{\psi} \to \overline{\psi} + \frac{i\omega_{\mu\nu}}{4}\overline{\psi}\sigma^{\mu\nu} \tag{14.47}$$

so that $\overline{\psi}^{\alpha}\psi_\alpha$ is an invariant.

Notice that

$$\begin{aligned}\overline{\psi}_L &= (\psi_L)^\dagger \gamma^0 \\ &= \left[\left(\frac{1+i\gamma^5}{2}\right)\psi\right]^\dagger \gamma^0 \\ &= \psi^\dagger \left(\frac{1-i\gamma^{5\dagger}}{2}\right)\gamma^0 \\ &= \overline{\psi}\gamma^0 \left(\frac{1-i\gamma^{5\dagger}}{2}\right)\gamma^0 \\ &= \overline{\psi} P_R \end{aligned} \tag{14.48}$$

so that $\overline{\psi}_L \psi_L = 0$ and the Weyl spinors are massless. Only when combined as in the (reducible) Dirac spinor do terms involving $\overline{\psi}_L \psi_R$ and $\overline{\psi}_R \psi_L$ become available as mass terms.

Charge Conjugation and Majorana Spinor

Now we must briefly discuss charge conjugation and the reality properties of spinors. If we have a Dirac equation minimally coupled (gauge) to electromagnetism we have

$$(i\,\partial\!\!\!/ + m - e\,A\!\!\!/)\psi = 0 \tag{14.49}$$

and we can easily establish that if we define the *charge conjugate* by

$$\psi_c = C\overline{\psi}^T \tag{14.50}$$

then we get

$$(i\,\partial\!\!\!/ + m - e\,A\!\!\!/)\psi_c = 0 \tag{14.51}$$

provided that

$$C(\gamma^\mu)^T C^{-1} = (-)\gamma^\mu. \tag{14.52}$$

Actually we should play safe in doubt and use the full indices

$$(\psi_c)_\alpha = C_{\alpha\beta}\overline{\psi}^\beta \tag{14.53}$$

$$\text{with} \quad C_{\alpha\beta}(\gamma^\mu)_\delta^{\ \beta}(C^{-1})^{\delta\epsilon} = (-)\,(\gamma^\mu)_\alpha^{\ \epsilon}. \tag{14.54}$$

At any rate we can establish that

$$
\begin{aligned}
C^T &= -C \\
(\gamma^5 C)^T &= -\gamma^5 C \quad \text{whereas} \quad (\gamma^\mu C)^T = -\gamma^\mu C \\
(\sigma^{\mu 5} C)^T &= -\sigma^{\mu 5} C \qquad\qquad (\sigma^{\mu\nu} C)^T = -\sigma^{\mu\nu} C.
\end{aligned}
\tag{14.55}
$$

In our original representation we can take a unitary C with $C^T = -C = C^\dagger = C^{-1}$ and $C^* = C$ as

$$
\begin{aligned}
C &= i\gamma^2\gamma^0 \\
&= \sigma^{20},
\end{aligned}
\tag{14.56}
$$

which looks like

$$
C = -i\begin{pmatrix} 0 & \sigma^2 \\ \sigma^2 & 0 \end{pmatrix}. \tag{14.57}
$$

In the Weyl representation this becomes

$$
\begin{aligned}
C &= \begin{pmatrix} -i\sigma^2 & 0 \\ 0 & i\sigma^2 \end{pmatrix} \\
&= \left(\begin{array}{cc|cc} 0 & -1 & & \\ 1 & 0 & & \\ \hline & & 0 & 1 \\ & & -1 & 0 \end{array}\right).
\end{aligned}
\tag{14.58}
$$

We see that

$$
\psi_c = \begin{pmatrix} i\sigma^2\psi_R^* \\ -i\sigma^2\psi_L^* \end{pmatrix}, \tag{14.59}
$$

so that the Weyl spinors form a *conjugate pair*. However, $(\psi_c)_c \equiv \psi_c$, so that reality can be imposed on the Dirac spinor. This is then known as a *Majorana spinor*. It then has the form

$$
\psi_M = \begin{pmatrix} \psi_L \\ -i\sigma^2\psi_L^* \end{pmatrix} \tag{14.60}
$$

and so $(\psi_M)_c \equiv \psi_M$.

Finally we return a few pages to rewrite the transformation as

$$
\psi_L \to \psi_L + i\omega^{0i}K^i\psi_L - \frac{i\omega^{ij}}{2}\varepsilon^{ijk}J^k\psi_L \tag{14.61}
$$

$$
\text{where} \quad J^i = \frac{1}{2}\sigma^i \quad \text{and} \quad k^i = \frac{i}{2}\sigma^i
$$

and we see that

$$
\begin{aligned}
[J^i, J^j] &= i\varepsilon^{ijk} J^k \\
[J^i, K^j] &= i\varepsilon^{ijk} K^k \\
[K^i, K^j] &= -i\varepsilon^{ijk} J^k
\end{aligned}
\tag{14.62}
$$

where it is the crucial minus sign in the last equation that shows we have $SL(2,C) \approx SO(1,3)$ and not $SU(2) \times SU(2)$.

A Notational Trick

A fairly standard description has been presented up to this point. Now it seems not to be widely recognized that in cases where there is a complex pair of spinors (as we have) then γ^5 can be taken along with C. More precisely, we shall switch our definition of C to a new one defined by

$$
\begin{aligned}
C \to -i\gamma^5 C &= \begin{pmatrix} i\sigma^2 & 0 \\ 0 & i\sigma^2 \end{pmatrix} \\
&= \left(\begin{array}{cc|cc} 0 & 1 & & \\ -1 & 0 & & \\ \hline & & 0 & 1 \\ & & -1 & 0 \end{array}\right)
\end{aligned}
\tag{14.63}
$$

in both representations. Also note that our new C has the matrix elements of $\varepsilon^{AB} \Rightarrow \varepsilon^{12} = 1 = -\varepsilon^{21}$. This works whether we are working with the left spinor or the right one. Moreover, the Majorana spinor now has the form

$$
\psi_M = \begin{pmatrix} \psi_L \\ i\sigma^2 \psi_L^* \end{pmatrix}
\tag{14.64}
$$

with the same matrix ε^{AB} form.

$SL(2, C)$ View

The literature switches between $SO(1,3)$ and $SL(2,C)$. Here is an $SL(2,C)$ view

$$
\psi'_\alpha = M_\alpha^{\ \beta} \psi_\beta \xrightarrow{*} \tilde{\psi}'^{T\alpha} = \tilde{\psi}^{T\beta} (M^{-1})_\beta^{\ \alpha}
\tag{14.65}
$$

where * is the definition of inverse by "raising" index

$$\overbrace{(\psi'^{\dagger})^{\dot\alpha}\{\equiv(\psi'_{\alpha})^*\}}^{\substack{\textit{Definition of}\\ \textit{complex dagger}\\ \Rightarrow\\ \textit{Raise and dot}}} = (M_{\alpha}{}^{\beta})^*\psi^{\dagger\dot\beta} \overset{Def^{\underline{n}}}{\equiv} \psi^{\dagger\dot\beta}(M^{\dagger})_{\dot\beta}^{\ \dot\alpha} \longrightarrow \overbrace{\tilde\psi'^{\dagger}_{\dot\alpha} = (M^{\dagger-1})_{\dot\alpha}^{\ \dot\beta}\tilde\psi^{\dagger}_{\dot\beta}}^{\substack{\textit{This is}\\ \textit{the Check}}}. \tag{14.66}$$

The dotted indices refer to the indices on the right-handed spinor; the undotted indices are on the left-handed spinor. The idea with a matrix group, given a representation ψ, is that other (possibly the same or related) representations are given by M^* (the conjugate) and M^{-1T} (the contragredient).

The important thing is to preserve the order

$$M(\theta_1, \theta_2) = M(\theta_1)M(\theta_2). \tag{14.67}$$

Of course it is highly convenient to have a matrix notation related to indices—even though the primitive idea is invariance by *saturation* (summing) of indices over upper and lower values of the same type (here dotted or undotted). Really $\tilde\psi$ is the (contragredient) representation, but we use $\tilde\psi^T$ for convenience. The index saturation and the matrix notation coexist happily, as long as we do not transpose at will. (Thus, for example, neither ψ^T nor $\tilde\psi$ have "indices" in the index saturation sense. There is no "raising" or "lowering"; there is no "metric.")

Note: In general the complex conjugate representation (and its contragredient one) have no relationship (other than conjugation) with the original one. This is why we use dotted indices. The dot comes from conjugation; the extra transposition is for later convenience.

Unitary Representations

(The idea is that for the subset of unitary matrices M we can drop the dots.) Unitarity is a property that survives the group law, so we can ask if $\tilde\psi$ transforms exactly as ψ^T under the subgroup—and up to a mixing of the components, of course. (This is why we threw a transposition into our definitions—for convenience at this point.)

So the question is: Is there a matrix $A_{\dot\alpha}^{\beta}$ such that $\overline{\psi}^{\alpha} \overset{Def}{\equiv} (\psi^{\dagger})^{\dot\beta} A_{\dot\beta}^{\alpha}$ transforms as $\tilde\psi^{\alpha}$? This is precisely what we used γ^0 for in defining $\overline{\psi}$. For the unitary subgroup ($SU(2)$ rotations) the matrix A is effectively unity—so the bar (or adjoint) of the spinor (or fundamental) is just the $\tilde\psi^T$ and $\overline{\psi}\psi$ is invariant.

Notice that our new choice of C is effectively $i\sigma^2$ in the subgroups, so we simply put

$$(i\sigma^2)_1^{\ 2} = \varepsilon^{12} = \varepsilon_{12} = \varepsilon^{\dot{1}\dot{2}} = \varepsilon_{\dot{1}\dot{2}} = 1. \tag{14.68}$$

We can proceed with some identities, an example of which is $\varepsilon^{\alpha\beta}\varepsilon_{\gamma\beta} = \delta^{\alpha}_{\ \gamma}$, and have nothing new to learn. However, we must be careful to "raise and lower" with matrix order in mind, for example,

$$\left.\begin{array}{lcl} V^{\alpha} & = & \varepsilon^{\alpha\beta}V_{\beta} \\ V_{\beta} & = & V_{\gamma}\varepsilon_{\gamma\beta} \end{array}\right\} \begin{array}{lcl} V^{\alpha} & \Rightarrow & \text{row} \\ V_{\beta} & \Rightarrow & \text{column} \end{array} \tag{14.69}$$

$$\begin{aligned} \text{therefore} \quad V^{\alpha} &= \varepsilon^{\alpha\beta}V_{\beta} = \varepsilon^{\alpha\beta}V^{\gamma}\varepsilon_{\gamma\beta} \\ &= V^{\gamma}\delta^{\alpha}_{\ \gamma} = V^{\alpha} \end{aligned} \tag{14.70}$$

works out correctly.

Supersymmetry: A First Look at the Simplest ($N = 1$) Case

It is a fundamental symmetry relating bosons and fermions. It merges Poincaré symmetry with internal symmetry in a way that does not violate the Coleman and Mandula "no go" theorem [7]. The theorem says that (apart from supersymmetry) we always get Poincaré $\otimes$ Internal, and that the internal is compact and semisimple. It was probably first descovered by Gol'fand and Likhtman [5], but really understood by Wess and Zumino [1,2,9].

In the simplest case we ask if we can extend the algebra of Poincaré $\otimes$ Internal by introducing supercharges Q_α and $\overline{Q}_{\dot{\alpha}}$, which have the *anticommutation* relations

$$\{Q_\alpha, \overline{Q}_{\dot{\beta}}\} = 2(\sigma^{\mu})_{\alpha\dot{\beta}}P_{\mu} \tag{14.71}$$

$$\{Q_\alpha, Q_\beta\} = 0 \tag{14.72}$$

and are fermionic. Notice that the closure is only onto the translations, which commute with themselves. This is essential, because the existence of a consistent algebra needs all possible generalized Jacobi identities to hold, and lots of zeros really help.

Here $\sigma^{\mu} \equiv \{1, \sigma^i\}$ makes the connection back between $SL(2, C)$ and $SO(1, 3)$. It turns out that we need simple commutators like

$$\left[Q_\alpha, P^{\mu}\right] = 0 \tag{14.73}$$

and the spinor nature of the supercharges requires

$$[Q_\alpha, M_{\mu\nu}] = \frac{1}{2}(\sigma_{\mu\nu})_\alpha^{\ \beta}Q_\beta. \tag{14.74}$$

The algebra is now closed. In this simple case, $N = 1$, there is just the one spinor Q_α and so no internal symmetry exists to label states, but see later.

Whenever the Q_α have a nontrivial effect on a state we get a change from fermion to boson, or vice versa. It is a theorem that *any supermultiplet has equal numbers of bosons and fermions.*

If N_F is the fermion number operator, then $(-1)^{N_F}$ has eigenvalue 1 for bosons, and eigenvalue –1 for fermions. Thus

$$(-1)^{N_F} Q_\alpha = -Q_\alpha (-1)^{N_F}. \tag{14.75}$$

Now

$$\begin{aligned} Tr\big[(-1)^{N_F}\{Q_\alpha, \overline{Q}_{\dot\beta}\}\big] &= Tr\big[(-1)^{N_F}(Q_\alpha \overline{Q}_{\dot\beta} + \overline{Q}_{\dot\beta} Q_\alpha)\big] \\ &= Tr\big[-Q_\alpha(-1)^{N_F}\overline{Q}_{\dot\beta} + Q_\alpha(-1)^{N_F}\overline{Q}_{\dot\beta}\big] \\ &= 0 \end{aligned} \tag{14.76}$$

since traces are cyclic for finite entries. But from the superalgebra we now have

$$2(\sigma^\mu)_{\alpha\dot\beta}\, Tr\big[(-1)^{N_F} P_\mu\big] = 0 \tag{14.77}$$

and P_μ is arbitrary so that

$$Tr[(-1)^{N_F}] = 0. \tag{14.78}$$

Thus there are equal numbers of fermions and bosons.

Actually this is stranger than it looks—it extends also from the on-mass-shell case here to field representations having to have matching numbers of components.

Massive Representations

The complete commuting set of observables includes p^μ, and if $p^2 = m^2$, we go to the rest frame as usual. What new things commute with the P^μ? Obviously the *supercharges*.

We define

$$a_\alpha = \frac{1}{\sqrt{2m}} Q_\alpha \tag{14.79}$$

$$\text{with} \quad (a_\alpha)^\dagger = \frac{1}{\sqrt{2m}} \overline{Q}_{\dot\alpha}. \tag{14.80}$$

Then

$$\{a_\alpha, (a_\beta)^\dagger\} = \delta_\alpha^{\ \beta} \tag{14.81}$$

$$\text{and} \quad \{a_\alpha, a_\beta\} = 0 = \{(a_\alpha)^\dagger, (a_\beta)^\dagger\} \tag{14.82}$$

where dots are dropped since we only have rotations in the complete commuting set of observables now.

We define a *Clifford vacuum* by

$$a_\alpha \Omega = 0 \tag{14.83}$$

noting that $p^2\Omega = m^2\Omega$ is not zero. The states arise by creating with $(a_\alpha)^\dagger$ on Ω. This soon stops because the $(a_\alpha)^\dagger$ anticommute.

All we get is

$$\begin{aligned} &\Omega \\ &(a_\alpha)^\dagger \Omega \\ \frac{1}{\sqrt{2}}(a_\alpha)^\dagger (a_\beta)^\dagger &= -\frac{1}{2\sqrt{2}} \varepsilon^{\alpha\beta} (a_\alpha)^\dagger (a_\beta)^\dagger \Omega. \end{aligned} \tag{14.84}$$

These are spin zero, spin $\frac{1}{2}$, and spin zero again. This is the *fundamental irreducible multiplet*. Note the match of spin zero bosons with the spin $\frac{1}{2}$ fermions.

We can also start with a vacuum Ω_j with $(2j+1)$ components of spin j. By addition of angular momentum we get

$$j \otimes [0 \oplus 1/2 \oplus 0] = j \oplus (j + 1/2) \oplus (j - 1/2) + j \tag{14.85}$$

as the multiplet. Figure 4.1 gives the multiplicities in tabular form.

The Clifford algebra structure reveals that we have an $SO(4)$ invariance group, which is isomorphic to $SU(2) \times SU(2)$. For more $Q_\alpha' s$—N replaces 1 here—we find $SO(4N) \supset SU(2) \times U\ Sp(2N)$ giving an "internal" symplectic $2N$ structure along with the $SU(2)$ spin.

Spin	Ω_0	$\Omega_{1/2}$	Ω_1	$\Omega_{3/2}$
0	2	1		
1/2	1	2	1	
1		1	2	1
3/2			1	2
2				1

FIGURE 14.1
The four vacuum spin components Ω_0, $\Omega_{\frac{1}{2}}$, Ω_1, $\Omega_{\frac{3}{2}}$ plotted against the spins 0, $\frac{1}{2}$, 1, $\frac{3}{2}$, 2.

Massless Representations

Now $p^2 = 0$, and we go to the $(E, 0, 0, E)$ frame and the Euclidean little group with helicity. Now the algebra of supercharges reads:

$$\{Q_\alpha, \overline{Q}_{\dot{\beta}}\} = 2\begin{pmatrix} 2E & 0 \\ 0 & 0 \end{pmatrix} \tag{14.86}$$

together with

$$\{Q_\alpha, Q_\beta\} = 0 = \{\overline{Q}_{\dot{\alpha}}, \overline{Q}_{\dot{\beta}}\}. \tag{14.87}$$

Defining

$$\begin{aligned} a &= \frac{1}{2\sqrt{E}} Q_i \\ a^\dagger &= \frac{1}{2\sqrt{E}} \overline{Q}_i \end{aligned} \tag{14.88}$$

we have

$$\begin{aligned} \{a, a^\dagger\} &= 1 \\ \{a, a\} &= 0 = \{a^\dagger, a^\dagger\}. \end{aligned} \tag{14.89}$$

But Q_2 and $Q_{\dot{2}}$ are totally anticommuting and must be represented by zero. We have a vacuum Ω_λ, of lowest helicity λ, abbreviated to "hel" in the table (see Figure 14.2), which is annihilated by $(a - iea\Omega_\lambda) = 0$. Then $a^\dagger\Omega_\lambda$ raises the helicity by $\frac{1}{2}$ but can only be used once. The massive representation splits into two massless ones. In CPT invariant schemes we usually need to double up all this. (But the cases for higher N sometimes are self-complete.)

λ hel	−2	−3/2	−1	−1/2	±0	+1/2	+1	+3/2
+2								1
+3/2							1	1
+1						1	1	
+1/2					1	1		
0				1	1			
−1/2			1	1				
−1		1	1					
−3/2	1	1						
−2	1							

FIGURE 14.2
The helicity pairs from $(-2, -\frac{3}{2})$ to $(3\frac{1}{2}, 2)$ plotted against the lowest helicity ("hel" in the table) of the vacuum $_\lambda$ of lowest helicity.

Superspace

I am going to introduce some new techniques that will make calculations easier. They are based on the idea of extending space–time into dimensions described by anticommuting coordinates. Just what this space is precisely does not seem to be well understood, although the formal manipulations seem to be consistent. To establish contact with the previous notation on Poincaré, and to provide a familiar example for reference, we shall have a review section.

Three-Dimensional Euclidean Space (Revisited)

The elements of the Hilbert space of quantum field theory are generated by the action of field-valued operators $\Phi(x)$ on a translationally invariant vacuum:

$$\begin{aligned} |\,\vec{x}\,\rangle &= \Phi(\vec{x}\,)\,|\,0\,\rangle \\ |\,\vec{x},\vec{x}\,'\,\rangle &= \Phi(\vec{x}\,)\Phi(\vec{x}\,')\,|\,0\,\rangle \\ &\vdots \end{aligned} \tag{14.90}$$

and translations of a state are generated by the energy-momentum operator:

$$\begin{aligned} |\,\vec{x}+\vec{a}\,\rangle &= \exp(-i\vec{a}\cdot\vec{P}\,)\,|\,\vec{x}\,\rangle \\ |\,\vec{x}+\vec{a},\vec{x}\,'+\vec{a}\,\rangle &= \exp(-i\vec{a}\cdot\vec{P}\,)\,|\,\vec{x},\vec{x}\,'\,\rangle, \end{aligned} \tag{14.91}$$

the displacement of a field is therefore given by

$$\begin{aligned} \Phi(\vec{x}-\vec{a}\,) &= e^{i\vec{a}\cdot\vec{P}}\,\Phi(\vec{x}\,)\,e^{-i\vec{a}\cdot\vec{P}} \\ &= \Phi(\vec{x}\,) + i[\vec{a}\cdot\vec{P},\Phi] + \cdots \end{aligned} \tag{14.92}$$

$$\text{and if}\quad [\Phi,\vec{P}] = -i\nabla\Phi \tag{14.93}$$

$$\begin{aligned} \text{then}\quad \delta\Phi &= \Phi'(\,\vec{x}\,) - \Phi(\vec{x}\,) \\ &= \Phi(\vec{x}-\vec{a}\,) - \Phi(\vec{x}\,) \\ &= -i[\Phi,\vec{a}\cdot\vec{P}] \\ &= (-i\vec{a}\,)\cdot(-i\nabla\Phi). \end{aligned} \tag{14.94}$$

In fact, using

$$e^{i\vec{a}\cdot\vec{P}}\,\Phi(\vec{x}\,)\,e^{-i\vec{a}\cdot\vec{P}} = \sum_{n=0}^{\infty}\frac{(i)^n}{n!}[\vec{a}\cdot\vec{P},[\vec{a}\cdot\vec{P},[\cdots[\vec{a}\cdot P,\Phi]]]_n \tag{14.95}$$

we have

$$\Phi'(\vec{x}) = \Phi(\vec{x} - \vec{a})$$
$$= \{\exp[-i\vec{a} \cdot (-i\nabla)]\}\Phi. \quad (14.96)$$

Now, suppose we start (as we shall in the supersymmetry case) simply with the algebra of the Euclidean group in three dimensions:

$$[M_{ij}, M_{kl}] = i(\delta_{ik}M_{jl} - \delta_{il}M_{jk} + \delta_{jl}M_{ik} - \delta_{jk}M_{il})$$
$$[M_{ij}, P_k] = i\delta_{ik}P_j - i\delta_{jk}P_i$$
$$[P_i, P_j] = 0 \quad (14.97)$$

where the first line is the algebra of the $SO(3)$ rotation subgroup. A group element can be specified by

$$g = \exp(-i)\left(a_i P_i + \frac{1}{2}\omega_{ij}M_{ij}\right) \quad (14.98)$$

where a_i and ω_{ij} are the (real) parameters. (There are equivalent alternatives, e.g., $\exp(-i\vec{a}' \cdot \vec{P})\exp(\frac{-i}{2}\omega_{ij}M_{ij})$, but this will not concern us.) If we define

$$L_i = \frac{1}{2}\varepsilon_{ijk}L_k$$
$$M_{ij} = \varepsilon_{ijk}L_k \quad (14.99)$$

$$\text{then} \quad [L_i, L_j] = i\varepsilon_{ijk}L_k$$
$$[L_i, P_j] = i\varepsilon_{ijk}P_k$$
$$[P_i, P_j] = 0 \quad (14.100)$$

is an equivalent form of the algebra, and

$$g = \exp(-i)(a_i P_i + \theta_i L_i) \quad (14.101)$$

where $\theta_i = \frac{1}{2}\varepsilon_{ijk}\omega_{jk}$ are the parameters now.

Now consider the three-dimensional coset space, which is the quotient of the Euclidean group by the $SO(3)$ rotation subgroup. We pick an origin in this space and put coordinates on its neighborhood by exponentiating the tangent space at that point; that is, we write a point in the coset space as

$$\mathcal{M}(\vec{x}) = \exp(-ix_i P_i) \quad (14.102)$$

where x_i are the coordinates. The *group action* is defined by left multiplication so that

$$g\ \exp(-ix_i P_i) = \exp(-ix'_i P_i)\exp(-i\eta_j L_j) \quad (14.103)$$

where η_j is whatever it has to be in terms of the coordinates and parameters, and x'_i are the coordinates of the transformed point.

The notation is general enough to handle more complicated problems later. The general theory [10] assures us that $x_i \to x_i'$ is a representation of the full group and is linear on the subgroup. In our simple case :

1. For translations where

$$g = \exp(-ia_i P_i) \tag{14.104}$$

$$\text{then} \quad g \exp(-ix_i P_i) = \exp(-i[x_i + a_i]P_i) \tag{14.105}$$

$$\text{and thus} \quad x_i' = x_i + a_i. \tag{14.106}$$

2. Again, for rotations, where

$$g = \exp(-i\theta_j L_j) \tag{14.107}$$

$$\begin{aligned}
\text{then} \quad g \exp(-ix_i P_i) &= \\
&= \exp(-i\vec{\theta}\cdot\vec{L}) \exp(-i\vec{x}\cdot\vec{P}) \exp(+i\vec{\theta}\cdot\vec{L}) \\
&\quad \times \exp(-i\vec{\theta}\cdot\vec{L}) \\
&= \exp\{-ix_i[\exp(-i\vec{\theta}\cdot\vec{L}) P_i \exp(+i\vec{\theta}\cdot\vec{L})]\} \\
&\quad \times \exp(-i\vec{\theta}\cdot\vec{L})
\end{aligned}$$

$$= \exp\left\{-ix_i\left(\sum_{n=0}^{\infty} \frac{(-i)^n}{n!}[\vec{\theta}\cdot\vec{L}, [\vec{\theta}\cdot\vec{L}, \cdots [\vec{\theta}\cdot\vec{L}, P_i]]]_n\right)\right\} \exp(-i\vec{\theta}\cdot\vec{L})$$

$$g \exp(-ix_i P_i) = \exp\{-ix_i(p_i - i\theta_j[L_j, P_i] + \cdots)\} \exp(-i\vec{\theta}\cdot\vec{L})$$

$$= \exp\{-i(x_i + \varepsilon_{ijk}\theta_j x_k + \cdots)P_i\} \exp(-i\vec{\theta}\cdot\vec{L}). \tag{14.108}$$

$$\text{Hence} \quad x_i' = x_i + \varepsilon_{ijk}\theta_j x_k + \cdots. \tag{14.109}$$

We have rediscovered the familiar translation and rotation induced transformations of the coordinates by the coset space method. Now we will use the idea of scalar fields to rediscover the representations of the generators of the algebra in terms of the coordinates and their derivatives. A scalar field has the property

$$\phi'(x') = \phi(x) \tag{14.110}$$

$$\text{or} \quad g\,\phi(x) = \phi(g^{-1}x). \tag{14.111}$$

1.

$$\begin{aligned} \text{So if} \quad g &= \exp(-ia_i P_i) \qquad (14.112) \\ \text{then} \quad \phi' &= \phi(x_i - a_i) \\ &= \phi(x) - ia_i(-i\partial_i)\phi + \cdots \\ &= \exp\{-ia_i(-i\partial_i)\}\phi \\ [\text{ or } \delta\phi &= \phi' - \phi = -ia_i(-i\partial_i)\phi\ ] \qquad (14.113) \\ \text{and we conclude} \quad \vec{P} &\Rightarrow -i\nabla. \qquad (14.114) \end{aligned}$$

2.

$$\text{Similarly, if} \quad g = \exp(-i\vec{\theta}\cdot\vec{L}) \qquad (14.115)$$

$$\begin{aligned} \text{then} \quad \phi' &= \phi(x_i - \varepsilon_{ijk}\theta_j x_k + \cdots) \\ &= \phi - i\theta_i\varepsilon_{ijk}x_j(-i\partial_k\phi\cdots \qquad (14.116) \end{aligned}$$

and we learn that

$$L_i \Rightarrow \varepsilon_{ijk}x_j(-i\partial_k) = (\vec{x}_n\vec{p})_i. \qquad (14.117)$$

$$\text{Notice that} \quad M_{ij} \Rightarrow \varepsilon_{ijk}L_k = -i(x_i\partial_j - x_j\partial_i). \qquad (14.118)$$

With these exercises under our belts, and the notation established, we turn to $N = 1$ superspace. This time we start with the *graded algebra*

$$\text{Poincaré Algebra}\left\{\begin{array}{l} \text{SO(1,3) Algebra}\left(\begin{array}{rcl} [M_{\mu\nu}, M_{\rho\sigma}] &=& ig_{\nu\rho}M_{\mu\sigma} + ig_{\mu\sigma}M_{\nu\rho} \\ && -ig_{\mu\rho}M_{\nu\sigma} - ig_{\nu\sigma}M_{\mu\rho} \\ [M_{\mu\nu}, P_\rho] &=& -ig_{\mu\nu}P_\nu + ig_{\nu\rho}P_\mu \end{array}\right) \\ \qquad [P_\mu, P_\nu] = 0 \end{array}\right\} \qquad (14.119)$$

$$\begin{aligned} &[Q_\alpha, P_\mu] = 0 \\ &[Q_\alpha, M_{\mu\nu}] = \frac{1}{2}(\sigma_{\mu\nu})_\alpha^{\ \beta}Q_\beta \\ &[\overline{Q}_{\dot\alpha}, M_{\mu\nu}] = -\frac{1}{2}(\overline{\sigma}^{\dot\beta}{}_{\dot\alpha}\overline{Q}_{\dot\beta} \\ &\text{with } \overline{\sigma}_{\mu\nu} = (\sigma_{\mu\nu})^\dagger \qquad (14.120) \end{aligned}$$

$$\begin{aligned} &\{Q_\alpha, \overline{Q}_{\dot\beta}\} = 2(\sigma^\mu)_{\alpha\dot\beta}P^\mu \\ &\{Q_\alpha, Q_\beta\} = 0. \qquad (14.121) \end{aligned}$$

In order to be able to copy the formalism we used in the previous case we introduce ξ^α, and $\overline{\xi}^{\dot\alpha} = (\xi^\alpha)^*$, as anticommuting (constant) parameters. The idea is that commutators of objects like $(\xi^\alpha Q_\alpha + \overline{Q}_{\dot\alpha}\overline{\xi}^{\dot\alpha})$ will then be specified by anticommutators of the Q_α and $\overline{Q}_{\dot\alpha}$, which are given in the algebra. Thus we can obtain group elements by exponentiation and work out products exactly as before. Explicitly, we have

$$\begin{aligned}\left[\xi^\alpha Q_\alpha + \overline{Q}_{\dot\alpha}\overline{\xi}^{\dot\alpha}, \eta^\beta Q_\beta + \overline{Q}_{\dot\beta}\overline{\eta}^{\dot\beta}\right] &= -\xi^\alpha\eta^\beta\{Q_\alpha, Q_\beta\} - \overline{\xi}^{\dot\alpha}\overline{\eta}^{\dot\beta}\{\overline{Q}_{\dot\alpha}, \overline{Q}_{\dot\beta}\} \\ &\quad + \xi^\alpha\overline{\eta}^{\dot\beta}\{Q_\alpha, \overline{Q}_{\dot\beta}\} + \overline{\xi}^{\dot\alpha}\eta^\beta\{\overline{Q}_{\dot\alpha}, Q_\beta\} \\ &= (\xi^\alpha\overline{\eta}^{\dot\beta} + \overline{\xi}^{\dot\beta}\eta^\alpha)2(\sigma^\mu)_{\alpha\dot\beta}P_\mu, \end{aligned} \tag{14.122}$$

which will be used shortly. Notice that if we want objects like $(\xi^\alpha Q_\alpha + \overline{Q}_{\dot\alpha}\overline{\xi}^{\dot\alpha})$ to be Hermitian then we must adopt the convention that *conjugation reverses the order of "fermionic" (i.e., odd Grassmann) factors without change of sign*. Notice also that the coefficient of P_μ in the last equation is an even element of a Grassmann algebra rather than an ordinary number. Thus the group parameters must share this property. This is a little disturbing, particularly as we are now going to parametrize a coset space to produce coordinates in superspace—our "ordinary" x^μ coordinates would seem to take values in the even part of the Grassmann algebra. We shall nevertheless differentiate without worrying further—all these formal manipulations seem to work out in a consistent manner.

Quite explicitly then we introduce anticommuting coordinates θ^α, and $\overline{\theta}^{\dot\alpha} = (\theta^\alpha)^*$, and extend our ordinary space–time described by x^μ to superspace described by $z^A = \{x^\mu; \theta^\alpha, \overline{\theta}^{\dot\alpha}\}$, by parametrizing the quotient space of the super Poincaré group by its $SO(1,3)$ Lorentz subgroup as

$$\mathcal{M} = \exp i\left(x^\mu p_\mu - \theta^\alpha Q_\alpha - \overline{Q}_{\dot\alpha}\overline{\theta}^{\dot\alpha}\right) \tag{14.123}$$

and then copy what we did in our example.

1. For space–time translations, when

$$g = \exp(ia^\mu P_\mu) \tag{14.124}$$

$$\text{then} \quad g\,\mathcal{M} = \exp i\left\{(x^\mu + a^\mu)P^\mu - \theta^\alpha Q_\alpha - \overline{Q}_{\dot\alpha}\overline{\theta}^{\dot\alpha}\right\} \tag{14.125}$$

$$\text{and thus} \quad \left\{\begin{matrix} x'^\mu & = & x^\mu + a^\mu \\ \theta'^\alpha = \theta^\alpha & , & \overline{\theta}'^{\dot\alpha} = \overline{\theta}^{\dot\alpha} \end{matrix}\right\}. \tag{14.126}$$

2. For Lorentz boosts and rotations, when

$$g = \exp\left(-\frac{i}{2}\omega^{\mu\nu}M_{\mu\nu}\right) \tag{14.127}$$

$$\begin{aligned}\text{then } g\mathcal{M} &= \exp\left(\frac{-i\vec{\omega}\cdot\vec{M}}{2}\right)\exp i(\vec{x}\cdot\vec{P} - \theta^\alpha Q_\alpha - \overline{Q}_{\dot\alpha}\overline{\theta}^{\dot\alpha}) \\ &\quad \exp\left(\frac{i\vec{\omega}\cdot\vec{M}}{2}\right)\exp\left(\frac{-i\vec{\omega}\cdot\vec{M}}{2}\right)\end{aligned} \tag{14.128}$$

so that

$$\mathcal{M}(z') = \exp\left\{ i \sum_{n=0}^{\infty} \frac{(-\frac{i}{2})^n}{n!} [\vec{\omega}\cdot\vec{M}, [\vec{\omega}\cdot\vec{M}, [\cdots[\vec{\omega}\cdot\vec{M}, x^\mu P_\mu - \theta^\alpha Q_\alpha - \overline{Q}_{\dot{\alpha}}\overline{\theta}^{\dot{\alpha}}]]]_n \right\}$$

$$= \exp i \left\{ x^\mu P_\mu - \theta^\alpha Q_\alpha - \overline{Q}_{\dot{\alpha}}\overline{\theta}^{\dot{\alpha}} - \frac{i\omega^{\rho\lambda}}{2}[M_{\rho\lambda}, x^\mu P_\mu - \theta^\alpha Q_\alpha - \overline{Q}_{\dot{\alpha}}\overline{\theta}^{\dot{\alpha}}] + \cdots \right\}$$

$$= \exp i \left\{ \begin{array}{l} x^\mu [P_\mu - \frac{1}{2} i\omega^{\rho\lambda}(-ig_{\rho\mu}P_\lambda + ig_{\lambda\mu}P_\rho)] \\ -\theta_\alpha [Q_\alpha + \frac{i}{2} i\omega^{\rho\lambda}\frac{1}{2}(\sigma_{\rho\lambda})_\alpha^{\ \beta} Q_\beta] \\ -[\overline{Q}_{\dot{\alpha}} - \frac{i}{2}\omega^{\rho\lambda}\frac{1}{2}(\overline{\sigma}_{\rho\lambda})^{\dot{\beta}}_{\ \dot{\alpha}} \overline{Q}_{\dot{\beta}}]\overline{\theta}^{\dot{\alpha}} \end{array} \right\}$$

$$= \exp i \left\{ \begin{array}{l} (x^\mu + \omega^\mu_{\ \nu} x^\nu) P_\mu \\ -[\theta^\alpha + \frac{i}{4}\omega^{\rho\lambda}(\sigma_{\rho\lambda})_\beta^{\ \alpha}]\, Q_\alpha \\ -\overline{Q}_{\dot{\alpha}}[\overline{\theta}^{\dot{\alpha}} - \frac{i}{4}\omega^{\rho\lambda}(\overline{\sigma}_{\rho\lambda})^{\dot{\alpha}}_{\ \dot{\beta}}\overline{\theta}^{\dot{\beta}}] \end{array} \right\} \tag{14.129}$$

and hence

$$\begin{aligned} x'^\mu &= x^\mu + \omega^\mu_{\ \nu} x^\nu \\ \theta'^\alpha &= \theta^\alpha + \frac{i}{4}\omega^{\rho\lambda}(\sigma_{\rho\lambda})_\beta^{\ \alpha}\theta^\beta \\ \overline{\theta}'^{\dot{\alpha}} &= \overline{\theta}^{\dot{\alpha}} - \frac{i}{4}\omega^{\rho\lambda}(\overline{\sigma}_{\rho\lambda})^{\dot{\alpha}}_{\ \dot{\beta}}\overline{\theta}^{\dot{\beta}}. \end{aligned} \tag{14.130}$$

3. Finally, if we have a pure super-transformation so that

$$g = \exp\{-i[\xi^\alpha Q_\alpha + \overline{Q}_{\dot{\alpha}}\overline{\xi}^{\dot{\alpha}}]\} \tag{14.131}$$

then $$g\,\mathcal{M} = \exp\{-i[\xi^\alpha Q_\alpha + \overline{Q}_{\dot{\alpha}}\overline{\xi}^{\dot{\alpha}}]\}\exp\{i[x^\mu P_\mu - \theta^\beta Q_\beta - \overline{Q}_{\dot{\beta}}\overline{\theta}^{\dot{\beta}}]\} \tag{14.132}$$

and we have to work this out explicitly using the Baker-Campbell-Hausdorff (BCH) identity

$$e^A e^B = e^{A+B+\frac{1}{2}[A,B]+c_1[A,[A,B]]+c_2[B,[A,B]]+\cdots}. \tag{14.133}$$

Fortunately only the first nontrivial term is needed (i.e., the terms with coefficients c_1, c_2, etc. are not needed). This is because the only nontrivial commutator involved gives P_μ, which then promptly commutes to zero with all the other P_μ, Q_α, and $\overline{Q}_{\dot{\alpha}}$. Explicitly we get

$$g\,\mathcal{M} = \exp i \left\{ \begin{array}{l} x^\mu P_\mu - (\theta^\alpha + \xi^\alpha) Q_\alpha - \overline{Q}_{\dot{\alpha}}(\overline{\theta}^{\dot{\alpha}} + \overline{\xi}^{\dot{\alpha}}) \\ -\frac{i}{2}[\xi^\alpha Q_\alpha + \overline{Q}_{\dot{\alpha}}\overline{\xi}^{\dot{\alpha}}, x^\mu P_\mu - \theta^\beta Q_\beta - \overline{Q}_{\dot{\beta}}\overline{\theta}^{\dot{\beta}}] \end{array} \right\}$$

$$= \exp i \left\{ \begin{array}{l} x^\mu P_\mu - (\theta^\alpha + \xi^\alpha) Q_\alpha - \overline{Q}_{\dot{\alpha}}(\overline{\theta}^{\dot{\alpha}} + \overline{\xi}^{\dot{\alpha}}) \\ +\frac{i}{2}(\xi^\alpha \overline{\theta}^{\dot{\beta}} + \overline{\xi}^{\dot{\beta}}\theta_\alpha) 2(\sigma^\mu)_{\alpha\dot{\beta}} P_\mu \end{array} \right\}$$

$$= \exp i \left\{ \begin{array}{l} [x^\mu + i(\xi^\alpha \overline{\theta}^{\dot{\beta}} + \overline{\xi}^{\dot{\beta}}\theta^\alpha)(\sigma^\mu)_{\alpha\dot{\beta}}] P_\mu \\ -(\theta^\alpha + \xi^\alpha) Q_\alpha - \overline{Q}_{\dot{\alpha}}(\overline{\theta}^{\dot{\alpha}} + \overline{\xi}^{\dot{\alpha}}) \end{array} \right\}. \quad (14.134)$$

Hence

$$\begin{aligned} x'^\mu &= x^\mu + i(\xi^\alpha \overline{\theta}^{\dot{\beta}} + \overline{\xi}^{\dot{\beta}}\theta^\alpha)(\sigma^\mu)_{\alpha\dot{\beta}} \\ \theta'^\alpha &= \theta^\alpha + \xi^\alpha \\ \overline{\theta}'^{\dot{\alpha}} &= \overline{\theta}^{\dot{\alpha}} + \overline{\xi}^{\dot{\alpha}} \end{aligned} \quad (14.135)$$

and we can see why people refer to these as *supertranslations*.

Now that we have our "coordinate transformations" we can use the idea of scalar field to derive the representations of our generators in terms of the coordinates. Taking the group elements in the same order as previously

1.

$$g = \exp\left(i a^\mu P_\mu\right) \quad (14.136)$$

$$\phi'(x,\theta,\overline{\theta}) = \phi(x^\mu - a^\mu, \theta, \overline{\theta}) \quad (14.137)$$

$$= \phi + i a^\mu i \partial_\mu \phi + \cdots . \quad (14.138)$$

So $\quad P_\mu \Rightarrow i\partial_\mu . \quad (14.139)$

$$g = \exp(-\frac{i}{2}\omega^{\mu\nu} M_{\mu\nu}) \quad (14.140)$$

$$\phi'(x,\theta,\overline{\theta}) = \phi\left(x^\mu - \omega^\mu{}_\nu x^\nu, \theta^\alpha - \frac{i}{4}\omega^{\rho\lambda}(\sigma_{\rho\lambda})_\beta{}^\alpha \theta^\beta, \overline{\theta}^{\dot{\alpha}} + \frac{i}{4}\omega^{\rho\lambda}(\overline{\sigma}_{\rho\lambda})^{\dot{\alpha}}{}_{\dot{\beta}}\overline{\theta}^{\dot{\beta}}\right) \quad (14.141)$$

$$= \phi - \frac{i}{2}\omega^{\mu\nu} \left\{ \begin{array}{c} i(x_\mu \partial_\nu - x_\nu \partial_\mu) \\ +\frac{1}{2}(\sigma_{\mu\nu})_\beta{}^\alpha \theta^\beta \partial_\alpha \\ -\frac{1}{2}(\overline{\sigma}_{\mu\nu})^{\dot{\alpha}}{}_{\dot{\beta}}\overline{\theta}^{\dot{\beta}}\overline{\partial}_{\dot{\alpha}} \end{array} \right\} \phi \quad (14.142)$$

where we have introduced

$$\partial_\alpha = \frac{\partial}{\partial\theta^\alpha} \text{ and } \bar{\partial}_{\dot\alpha} = \frac{\partial}{\partial\overline{\theta}^{\dot\alpha}} \tag{14.143}$$

so defined such that

$$\partial_\alpha\theta^\beta = \delta_\alpha^{\ \beta} \text{ and } \bar{\partial}_{\dot\alpha}\overline{\theta}^{\dot\beta} = \delta_{\dot\alpha}^{\ \dot\beta}. \tag{14.144}$$

(Note that ∂_α, and $\bar{\partial}_{\dot\alpha}$, anticommute with other Fermi-like objects. Also, don't confuse

$$\partial_\alpha = \frac{\partial}{\partial\theta^\alpha} \quad \text{with} \quad \partial_\mu = \frac{\partial}{\partial x^\mu} \tag{14.145}$$

and similarly

$$\theta_i \quad \text{with} \quad \theta_\alpha. \tag{14.146}$$

In practice this "confusion" does not seem to become a major problem.) We find, therefore,

$$M_{\mu\nu} \Rightarrow i(x_\mu\partial_\nu - x_\nu\partial_\mu) + \frac{1}{2}(\sigma_{\mu\nu})_\beta^{\ \alpha}\theta^\beta\partial_\alpha - \frac{1}{2}(\overline{\sigma}_{\mu\nu})^{\dot\alpha}_{\ \dot\beta}\overline{\theta}^{\dot\beta}\bar{\partial}_{\dot\alpha}. \tag{14.147}$$

2. Finally, if

$$g = \exp\{-i(\xi^\alpha Q_\alpha + \overline{Q}_{\dot\alpha}\overline{\xi}^{\dot\alpha})\} \tag{14.148}$$

then

$$\phi' = \phi[x^\mu - i(\xi^\alpha\overline{\theta}^{\dot\beta} + \overline{\xi}^{\dot\beta}\theta^\alpha)(\sigma)_{\alpha\dot\beta};\theta^\alpha - \xi^\alpha, \overline{\theta}^{\dot\alpha} - \overline{\xi}^{\dot\alpha}] \tag{14.149}$$

$$= \phi - i\left\{\begin{array}{l}\xi^\alpha(-i\partial_\alpha + \overline{\theta}^{\dot\beta}(\sigma^\mu)_{\alpha\dot\beta}\partial_\mu) \\ +(i\bar{\partial}_{\dot\alpha} - \theta^\beta(\sigma^\mu)_{\beta\dot\alpha}\partial_\mu)\overline{\xi}^{\dot\alpha}\end{array}\right\}\phi + \cdots. \tag{14.150}$$

(Watch the anticommutator signs; also ξ^α are constant here.) Hence

$$Q_\alpha \Rightarrow -i\partial_\alpha + \overline{\theta}^{\dot\beta}(\sigma^\mu)_{\alpha\dot\beta}\partial_\mu \equiv -i\partial_\alpha + \overline{\theta}^{\dot\beta}\partial_{\alpha\dot\beta} \tag{14.151}$$

$$\overline{Q}_{\dot\alpha} \Rightarrow -i\bar{\partial}_{\dot\alpha} - \theta^\beta(\sigma^\mu)_{\beta\dot\alpha}\partial_\mu \equiv +i\bar{\partial}_{\dot\alpha} + \theta^\beta\partial_{\beta\dot\alpha}. \tag{14.152}$$

Covariant Derivative Operators from Right Action

We found a representation of the group action by multiplying the group element from the left into the coset space. We can also get a (conjugate) action by multiplying from the right. Because we like to keep track of signs, we will multiply from the right by $(g)^{-1}$; then since $(g_2g_1)^{-1} = (g_1)^{-1}(g_2)^{-1}$ the composition law and the subgroup action are identical.

Now, when the commutators (anticommuting parameters are used when needed) of the coset space generators close onto a subset, which commute to zero with the original set, then the situation is very simple and we can easily find operators (useful ones) that act from the left on the coset space to give the right multiplication action. This is easier than it sounds. Here

$$[P_\mu, P_\nu] = 0 \tag{14.153}$$

$$[P_\mu, Q_\alpha] = 0 \tag{14.154}$$

$$\{Q_\alpha, Q_\beta\} = 0 \tag{14.155}$$

$$\{Q_\alpha, \overline{Q}_{\dot\beta}\} = 2(\sigma^\mu)_{\alpha\dot\beta} P_\mu. \tag{14.156}$$

Hence, if we define our action as described, we have

$$\begin{aligned}
&\mathcal{M}(x,\theta,\overline{\theta})g^{-1}(0,\xi,\overline{\xi})\\
&\qquad = \exp i(x^\mu P_\mu - \theta^\alpha Q_\alpha - \overline{Q}_{\dot\alpha}\overline{\theta}^{\dot\alpha}) \exp i(\xi^\alpha Q_\alpha + \overline{Q}_{\dot\alpha}\overline{\xi}^{\dot\alpha})\\
&\qquad = e^{i(x,\theta,\overline{\theta})} e^{i(0,-\xi,-\overline{\xi})}\\
&\qquad = e^{i(x,\theta,\overline{\theta})} e^{i(0,-\xi,-\overline{\xi})} e^{-i(x,\theta,\overline{\theta})} e^{i(x,\theta,\overline{\theta})}\\
&\qquad = \exp\left\{\sum_{n=0}^{\infty} \frac{(-i)^n}{n!}[(\theta,\overline{\theta}),[(\theta,\overline{\theta}),[\cdots[(\theta,\overline{\theta}),i(\xi,\overline{\xi})]]]_n\right\} e^{i(x,\theta,\overline{\theta})}\\
&\qquad = \exp\{i(\xi,\overline{\xi}) - i[(\theta,\overline{\theta}),i(\xi,\overline{\xi})] + \text{zero}\} e^{i(x,\theta,\overline{\theta})}\\
&\qquad = \exp\left\{i(\xi^\alpha Q_\alpha + \overline{Q}_{\dot\alpha}\overline{\xi}^{\dot\alpha}) - (\xi^\alpha\overline{\theta}^{\dot\alpha} + \overline{\xi}^{\dot\alpha}\theta^\alpha)2(\sigma^\mu)_{\alpha\dot\alpha}P_\mu\right\} e^{i(x,\theta,\overline{\theta})}\\
&\qquad = \exp\left\{i(\xi^\alpha \hat{Q}_\alpha + \hat{\overline{Q}}_{\dot\alpha}\overline{\xi}^{\dot\alpha})\right\} \exp\{i(x^\mu P_\mu - \theta^\alpha Q_\alpha - \overline{Q}_{\dot\alpha}\overline{\theta}^{\dot\alpha})\}
\end{aligned} \tag{14.157}$$

where we have used

$$\hat{Q}_\alpha = Q_\alpha + 2i\overline{\theta}^{\dot\alpha}(\sigma^\mu)_{\alpha\dot\alpha}P_\mu \tag{14.158}$$

$$\hat{\overline{Q}}_{\dot\alpha} = \overline{Q}_{\dot\alpha} - 2i\theta^\alpha(\sigma^\mu)_{\alpha\dot\alpha}P_\mu. \tag{14.159}$$

Now, although this came out as expected, we may wonder where all this is going. Just see what associativity of the group law gives; ξ and η are parameters

$$
\begin{aligned}
e^{i(\xi\hat{Q}+\overline{\hat{Q}}\overline{\xi})} & \left\{e^{-i(\eta Q+\overline{Q}\overline{\eta})}\mathcal{M}(z)\right\} \\
&= e^{i(\xi\hat{Q}+\overline{\hat{Q}}\overline{\xi})}\mathcal{M}(z') \\
&= \mathcal{M}(z')e^{i(\xi Q+\overline{Q}\overline{\xi})} \\
&= \left\{e^{-i(\eta Q+\overline{Q}\overline{\eta})}\mathcal{M}(z)\right\}e^{i(\xi Q+\overline{Q}\overline{\xi})} \\
&= e^{-i(\eta Q+\overline{Q}\overline{\eta})}\left\{\mathcal{M}(z)e^{i(\xi Q+\overline{Q}\overline{\xi})}\right\} \\
&= e^{-i(\eta Q+\overline{Q}\overline{\eta})}\left\{e^{i(\xi Q+\overline{Q}\overline{\xi})}\mathcal{M}(z)\right\}. \qquad (14.160)
\end{aligned}
$$

Now compare this to the start, and notice that η and ξ are *arbitrary* $\Rightarrow$

$$\{\hat{Q}, Q\} = 0 = \{\overline{\hat{Q}}, Q\}. \qquad (14.161)$$

Hence

$$D_\alpha \equiv +i\hat{Q}_\alpha = \partial_\alpha - i\overline{\theta}^{\dot{\beta}}(\sigma^\mu)_{\alpha\dot{\beta}}\partial_\mu = \partial_\alpha - i\overline{\theta}^{\dot{\beta}}\partial_{\alpha\dot{\beta}} \qquad (14.162)$$

$$\text{and} \quad \overline{D}_{\dot{\alpha}} \equiv -i\overline{\hat{Q}}_\alpha = \overline{\partial}_{\dot{\alpha}} - i\theta^\beta(\sigma^\mu)_{\beta\dot{\alpha}}\partial_\mu = \overline{\partial}_{\dot{\alpha}} - i\theta^\beta\partial_{\beta\dot{\alpha}} \qquad (14.163)$$

are covariant derivative operators. Note that $\partial_\mu = \partial/\partial x^\mu$ was already known to be fine, since $[P_\mu, Q_\alpha] = 0$.

Of course, we could have discovered this directly but the treatment here reveals the underlying motivation, and will be useful as a model for other ideas (in compactification) coming later.

We finally compute (a trivial excercise) directly that

$$
\begin{aligned}
\{D_\alpha, \overline{D}_{\dot{\beta}}\} &= (-)2(\sigma^\mu)_{\alpha\dot{\beta}}(i\partial_\mu) && (14.164)\\
&= (-)2i\partial_{\alpha\dot{\beta}} && (14.165)\\
&= (-)2(\sigma^\mu)_{\alpha\dot{\beta}}P_\mu. && (14.166)
\end{aligned}
$$

This is frequently useful and also reveals the freedom of the curious representation of Q_α used by Wess and Zumino [9].

Superfields

What are these scalar fields $\Phi(x, \theta, \overline{\theta})$ in this curious space? We have implied their existence, but done little else with them. The basic point is that powers of odd (fermionic) objects eventually terminate (nilpotency).

For example,

$$\theta_\alpha \theta_\beta = (-)\frac{1}{2}\varepsilon_{\alpha\beta}\theta^2 \qquad (14.167)$$
$$\text{with} \quad \theta^2 = \theta^\alpha \theta_\alpha$$

is fine, but

$$\theta_\alpha \theta_\beta \theta_\gamma \equiv 0 \qquad (14.168)$$

by antisymmetry. Similarly

$$\overline{\theta}^{\dot\alpha} \overline{\theta}^{\dot\beta} = \frac{1}{2}\varepsilon^{\dot\alpha\dot\beta}\overline{\theta}^2 \qquad (14.169)$$
$$\text{with} \quad \overline{\theta}^2 = \overline{\theta}_{\dot\alpha}\overline{\theta}^{\dot\alpha}$$

is as far as we can go again. So superfields $\Phi(x, \theta, \overline{\theta})$ are (complex numbers, Grassmann valued) functions defined on (subsets of) $\mathcal{M}$, which must be understood in terms of their power series expansion in θ^α and $\overline{\theta}^{\dot\alpha}$

$$\begin{aligned}\Phi(x, \theta, \overline{\theta}) =& f(x) + \theta^\alpha \phi_\alpha(x) + \overline{\theta}_{\dot\alpha}\chi^{\dot\alpha}(x) + \theta^2 m(x) + \overline{\theta}^2 n(x) \\ &+ \theta_\alpha \overline{\theta}_{\dot\beta} V^{\alpha\dot\beta}(x) + \theta^2 \overline{\theta}_{\dot\alpha}\lambda^{\dot\alpha}(x) + \overline{\theta}^2 \theta^\alpha \psi_\alpha(x) + \theta^2 \overline{\theta}^2 D(x) \end{aligned} \qquad (14.170)$$

and we notice that spins up to unity can occur in a single superfield (without Lorentz indices). Note also that our definition of differentiation with respect to spinor coordinates is

$$\Phi(x; \theta + \delta\theta, \overline{\theta} + \delta\overline{\theta}) = \Phi(x; \theta, \overline{\theta}) + \left(\delta\theta^\alpha \partial_\alpha + \delta\overline{\theta}^{\dot\alpha}\overline{\partial}_{\dot\alpha}\right)\Phi(x; \theta, \overline{\theta}) \qquad (14.171)$$

and that there is a graded Leibnitz rule

$$\partial_\alpha(AB) = (\partial_\alpha A)B + (-1)^{\{A\}} A(\partial_\alpha B) \qquad (14.172)$$
$$\text{where} \quad \{A\} = \begin{cases} 0 & \text{if A is bosonic} \\ -1 & \text{if A is fermionic} \end{cases}.$$

Supertransformations

We can now rederive results previously obtained. Consider the full field

$$\Phi = f(x) + \cdots\cdots + \theta^2\overline{\theta}^2 D(x) \tag{14.173}$$

$$\text{where} \quad \delta_\xi \Phi = (-)i(\xi^\alpha Q_\alpha + \overline{Q}_{\dot\alpha}\overline{\xi}^{\dot\alpha})\Phi$$

$$= (-)\left\{ \begin{array}{c} \xi^\alpha\left[\partial_\alpha + i\overline{\theta}^{\dot\beta}(\sigma^\mu)_{\alpha\dot\beta}\partial_\mu\right] \\ +\overline{\xi}^{\dot\alpha}\left[\overline{\partial}_{\dot\alpha} + i\theta^\beta(\sigma^\mu)_{\beta\dot\alpha}\partial_\mu\right] \end{array}\right\}\Phi$$

and work this out to equate to

$$\delta_\xi f(x) + \cdots\cdots + \theta^2\overline{\theta}^2(\delta_{\overline{\xi}} D(x)) \tag{14.174}$$

and thus find the changes in the component fields. We will just work out one piece

$$\theta^2\overline{\theta}^2\delta_\xi D(x) = (-)\xi^\alpha i\overline{\theta}^{\dot\beta}(\sigma^\mu)_{\alpha\dot\beta}\partial_\mu\{\theta^2\overline{\theta}_{\dot\delta}\lambda^{\dot\delta}(x)\}$$
$$- \overline{\xi}^{\dot\alpha} i\theta^\beta(\sigma^\mu)_{\beta\dot\alpha}\partial_\mu\{\overline{\theta}^2\theta^\varepsilon\psi_\varepsilon(x)\}. \tag{14.175}$$

$$\text{Now} \quad \theta^\alpha\theta^\beta = (-)\frac{1}{2}\varepsilon^{\alpha\beta}\theta^2$$
$$\text{with} \quad \theta^2 = \theta^\alpha\theta_\alpha$$
$$\text{and similarly} \quad \overline{\theta}^{\dot\alpha}\overline{\theta}^{\dot\beta} = (+)\frac{1}{2}\varepsilon^{\dot\alpha\dot\beta}\overline{\theta}^2$$
$$\text{with} \quad \overline{\theta}^2 = \overline{\theta}_{\dot\alpha}\overline{\theta}^{\dot\alpha}.$$

Apologies to other authors, but the notation here allows our conjugation of equations—so now $\varepsilon_{\dot\alpha\dot\beta} = \overline{\varepsilon_{\alpha\beta}}(\equiv \varepsilon_{\alpha\beta}$ numerically), and dotted indices are raised and lowered exactly as undotted ones.

Hence

$$\theta^2\overline{\theta}^2\delta_\xi D(x) = \xi^\alpha\frac{i}{2}(\sigma^\mu)_{\alpha\dot\beta}\theta^2\overline{\theta}^2\partial_\mu\lambda^{\dot\beta}(x) + \overline{\xi}^{\dot\alpha}\frac{i}{2}(\sigma^\mu)_{\beta\dot\alpha}\varepsilon^{\beta\varepsilon}\theta^2\overline{\theta}^2\partial^\mu\psi_\varepsilon(x)$$
$$= \frac{i}{2}\theta^2\overline{\theta}^2(\sigma^\mu)_{\alpha\dot\beta}\{\xi^\alpha\partial_\mu\lambda^{\dot\beta} + \overline{\xi}^{\dot\beta}\partial_\mu\psi^\alpha\} \tag{14.176}$$
$$\text{or} \quad \delta_\xi D(x) = \frac{i}{2}\partial_{\alpha\dot\beta}\{\xi^\alpha\lambda^{\dot\beta} + \overline{\xi}^{\dot\beta}\psi^\alpha\} \tag{14.177}$$

since the parameters are constant. Notice that this is a total derivative.

Notes

1. The product of two superfields is (trivially) a superfield, so that superfields are like the tensors of supersymmetry.

2. In general, a superfield will be reducible. We find the irreducible representations by imposing constraints that do not yield equations of motion in four-dimensional space–time. These can be reality conditions, or $D_\alpha \Phi = 0$ type conditions, as we shall see. There is so much new notation to confuse us that we had better make contact with something simple, then go on later to see its full beauty and power.

The Chiral Scalar Multiplet

We shall show that the condition

$$\overline{D}_{\dot{\alpha}} \Phi = 0 \tag{14.178}$$

gives exactly an irreducible multiplet. Recall that

$$\overline{D}_{\dot{\alpha}} = \bar{\partial}_{\dot{\alpha}} + i\theta^{\beta}(\sigma^{\mu})_{\beta\dot{\alpha}}\partial_{\mu} \tag{14.179}$$

$$\text{so that} \quad \overline{D}_{\dot{\alpha}}\theta^{\beta} = 0 \tag{14.180}$$

$$\text{and} \quad \overline{D}_{\dot{\alpha}} y^{\mu} = 0 \tag{14.181}$$

$$\text{where} \quad y^{\mu} = x^{\mu} - i(\sigma^{\mu})_{\rho\dot{\lambda}}\theta^{\rho}\overline{\theta}^{\dot{\lambda}}$$

follows trivially. This means that any function $\Phi(y, \theta)$ is a solution of the constraint. We can expand this as

$$\Phi = A(y) - 2\theta^{\alpha}\psi_{\alpha}(y) - \theta^{2}F(y) \tag{14.182}$$

with the notation previously introduced; the factors and signs are designed to give exact contact with Wess and Zumino. Now it would be convenient to express this back in terms of x, θ, and $\overline{\theta}$.

Notice that

$$\{-i\theta^{\alpha}\overline{\theta}^{\dot{\beta}}(\sigma^{\nu})_{\alpha\dot{\beta}}\partial_{\nu}\}x^{\mu} = -i\theta^{\alpha}\overline{\theta}^{\dot{\beta}}(\sigma^{\mu})_{\alpha\dot{\beta}} \tag{14.183}$$

$$\begin{aligned}\text{so that} \quad \Phi(y,\theta) &= \exp\left(-i\theta^{\alpha}\overline{\theta}^{\dot{\beta}}\partial_{\alpha\dot{\beta}}\right)\left\{A(x) - 2\theta^{\alpha}\psi_{\alpha}(x) - \theta^{2}F(x)\right\} \\ &= \left(1 - i\theta^{\alpha}\overline{\theta}^{\dot{\beta}}\partial_{\alpha\dot{\beta}} - \frac{1}{2}\theta^{\alpha}\overline{\theta}^{\dot{\beta}}\partial_{\alpha\dot{\beta}}\theta^{\gamma}\overline{\theta}^{\dot{\delta}}\partial_{\gamma\dot{\delta}}\right)\{A - 2\theta^{\rho}\psi_{\rho} - \theta^{2}F\}\end{aligned} \tag{14.184}$$

$$= \left.\begin{matrix} A(x) - 2\theta^{\rho}\psi_{\rho}(x) - \theta^{2}F(x) - i\theta^{\alpha}\overline{\theta}^{\dot{\beta}}\partial_{\alpha\dot{\beta}}A(x) \\ +i\theta^{2}\overline{\theta}^{\dot{\beta}}\partial_{\alpha\dot{\beta}}\psi^{\alpha}(x) - \frac{1}{4}\theta^{2}\overline{\theta}^{2}\Box A(x)\end{matrix}\right\} \tag{14.185}$$

where we have used $\partial_{\alpha\dot{\beta}}\partial^{\alpha\dot{\beta}} = 2\Box$. Clearly this has the same content we had previously. Now we must check that the transformation rules are the same.

We just work out a supercharge

$$\delta_\xi \Phi = (-)[\xi^\alpha Q_\alpha + \overline{Q}_{\dot\alpha}\overline{\xi}^{\dot\alpha}]\Phi \tag{14.186}$$

$$= (-)\left\{\begin{array}{l}\xi^\alpha\left(\partial_\alpha + i\overline{\theta}^{\dot\beta}\partial_{\alpha\dot\beta}\right) \\ \\ +\overline{\xi}^{\dot\alpha}(\overline{\partial}_{\dot\alpha} + i\theta^\beta\partial_{\beta\dot\alpha})\end{array}\right\}\{Eq^n\ (14.183)\ \text{above}\} \tag{14.187}$$

$$= 2\xi^\alpha\psi_\alpha(x) + 2\xi^\alpha\theta_\alpha F(x) - i\overline{\xi}^{\dot\alpha}\theta^\rho\partial_{\rho\dot\alpha}A(x)$$

$$- i\overline{\xi}^{\dot\alpha}\theta^2\partial_{\rho\dot\alpha}\psi^\rho(x) - \overline{\xi}^{\dot\alpha}i\theta^\beta\partial_{\beta\dot\alpha}\left[A(x) - 2\theta^\rho\psi_\rho(x)\right] \tag{14.188}$$

$$+ \text{terms in } \overline{\theta}. \tag{14.189}$$

(We used $\partial_\alpha\theta^2 = 2\theta_\alpha$.)

$$= 2\xi^\alpha\psi_\alpha(x) - 2\theta^\alpha\left[-i\overline{\xi}^{\dot\alpha}\partial_{\alpha\dot\alpha}A(x) - \xi_\alpha F(x)\right]$$

$$- \theta^2 2i\overline{\xi}^{\dot\alpha}\partial_{\rho\dot\alpha}\psi^\rho(x) + \text{terms in } \overline{\theta}. \tag{14.190}$$

Hence

$$\delta_\xi A(x) = 2\xi^\alpha\psi_\alpha(x) \tag{14.191}$$

$$\delta_\xi\psi_\alpha(x) = (-)i\overline{\xi}^{\dot\alpha}\partial_{\alpha\dot\alpha}A(x) - \xi_\alpha F(x) \tag{14.192}$$

$$\delta_\xi F(x) = 2i\overline{\xi}^{\dot\alpha}\partial_{\rho\dot\alpha}\psi^\rho(x) \tag{14.193}$$

and we have confirmed the Wess and Zumino result. Notice the familiar total divergence in $\delta_\xi F$.

Superspace Methods

We are now going to exhibit some of the power of these methods. First, we record some very useful identities that we can check directly and quickly:

$$\{D_\alpha, D_\beta\} = 0 \tag{14.194}$$

$$\left[D_\alpha, \overline{D}^2\right] = (-)4i\partial_{\alpha\dot\alpha}\overline{D}^{\dot\alpha} \tag{14.195}$$

$$\{D_\alpha, \overline{D}_{\dot\beta}\} = (-)2i\partial_{\alpha\dot\beta} \tag{14.196}$$

$$\left[\overline{D}_{\dot\alpha}, D^2\right] = 4i\partial_{\alpha\dot\alpha}D^\alpha \tag{14.197}$$

$$\{\overline{D}_{\dot\alpha}, \overline{D}_{\dot\beta}\} = 0 \tag{14.198}$$

$$\left[D^2, \overline{D}^2\right] = 4i\partial_{\alpha\dot\alpha}\left[\overline{D}^{\dot\alpha}, D^\alpha\right] \tag{14.199}$$

$$D^2\overline{D}^2D^2 = (-)16\Box D^2 \tag{14.200}$$

$$\overline{D}^2D^2\overline{D}^2 = (-)16\Box\overline{D}^2. \tag{14.201}$$

On a chiral superfield, when

$$\overline{D}_{\dot{\alpha}}\Phi = 0, \tag{14.202}$$

$$\overline{D}_{\dot{\alpha}} D^2\Phi = 4i\partial_{\alpha\dot{\alpha}} D^{\alpha}\Phi \tag{14.203}$$

$$\text{and} \quad \overline{D}^2 D^2\Phi = (-)16\Box\Phi. \tag{14.204}$$

Covariant Definition of Component Fields

We know that $\partial_\alpha = \partial/\partial\theta^{\alpha}$ is not covariant. We can improve on this by using D_α, and can satisfy differential constraints automatically. For example, take our chiral field with $\overline{D}_{\dot{\alpha}} = 0$. Define (with previous notation in mind)

$$\Phi|_{\theta=0=\overline{\theta}} = A \tag{14.205}$$

$$D_\alpha\Phi|_0 = (-)2\psi_\alpha \tag{14.206}$$

$$D^2\Phi\Big|_0 = 4F. \tag{14.207}$$

Supercharges Revisited

We can work out the supercharge much more easily now. Notice that

$$D_\alpha|_0 = i\hat{Q}_\alpha\Big|_0 \tag{14.208}$$

$$= Q_\alpha|_0 \tag{14.209}$$

$$\text{and} \quad \overline{D}_{\dot{\alpha}}|_0 = (-)i\hat{\overline{Q}}_{\dot{\alpha}}|_0 \tag{14.210}$$

$$= -i\overline{Q}_{\dot{\alpha}}|_0, \tag{14.211}$$

which allows us to use our identities. (Warning: Naturally $D_\alpha\overline{Q}_{\dot{\beta}}|_0 \neq D_\alpha i\overline{D}_{\dot{\beta}}|_0$, so we must take Q and $\overline{Q}$ to the left of all derivatives before switching to D and $\overline{D}$.) Then, for example,

$$\delta_\xi A = \delta\Phi|_0 \tag{14.212}$$

$$= (-i)(\xi^{\alpha}Q_\alpha + \overline{Q}_{\dot{\alpha}}\overline{\xi}^{\dot{\alpha}})\Phi\Big|_0 \tag{14.213}$$

$$= (-\xi^{\alpha}D_\alpha + \overline{D}_{\dot{\alpha}}\overline{\xi}^{\dot{\alpha}})\Phi\Big|_0 \tag{14.214}$$

$$= -\xi^{\alpha}D_\alpha\Phi|_0 \tag{14.215}$$

$$= 2\xi^{\alpha}\psi_\alpha(x), \tag{14.216}$$

which is the Wess and Zumino result yet again.

$$\text{Again}\quad \delta_\xi \psi = (-)\frac{1}{2} D_\alpha \delta_\xi \Phi\Big|_0 \tag{14.217}$$

$$= D_\alpha(-i)\{\xi^\beta Q_\beta + \overline{Q}_{\dot\beta}\overline{\xi}^{\dot\beta}\}\Phi\Big|_0 \tag{14.218}$$

$$= \frac{i}{2}[\xi^\beta Q_\beta + \overline{Q}_{\dot\beta}\overline{\xi}^{\dot\beta}]D_\alpha\Phi\Big|_0 \quad \text{(this step is essential)} \tag{14.219}$$

$$= \frac{1}{2}[\xi^\beta D_\beta - \overline{D}_{\dot\beta}\overline{\xi}^{\dot\beta}]D_\alpha\Phi\Big|_0 \tag{14.220}$$

$$= \frac{1}{2}\xi^\beta D_\beta D_\alpha \Phi\Big|_0 + \frac{1}{2}\overline{\xi}^{\dot\beta}\{\overline{D}_{\dot\beta}, D_\alpha\}\Phi\Big|_0 \tag{14.221}$$

$$= \frac{1}{2}\xi^\beta\left(-\frac{1}{2}\right)\varepsilon_{\beta\alpha}D^2\Phi\Big|_0 - i\overline{\xi}^{\dot\beta}\partial_{\alpha\dot\beta}\Phi\Big|_0 \tag{14.222}$$

$$= -\xi_\alpha F - i\overline{\xi}^{\dot\beta}\partial_{\alpha\dot\beta}A \tag{14.223}$$

and we confirm the previous result.

$$\text{Finally}\quad \delta_\xi F = \frac{1}{4}D^2\delta_\xi\Phi\Big|_0 \tag{14.224}$$

$$= \frac{1}{4}\delta_\xi D^2\Phi\Big|_0 \tag{14.225}$$

$$= \frac{1}{4}(-i)[\xi^\alpha Q_\alpha + \overline{Q}_{\dot\alpha}\overline{\xi}^{\dot\alpha}]D^2\Phi\Big|_0 \tag{14.226}$$

$$= (-)\frac{1}{4}\overline{\xi}^{\dot\beta}\overline{D}_{\dot\beta}D^2\Phi\Big|_0 \quad \ldots \text{ since } D^3 \equiv 0 \tag{14.227}$$

$$= (-)\frac{1}{4}\overline{\xi}^{\dot\beta}\, 4i\partial_{\beta\dot\beta}D^\beta\Phi\Big|_0 \quad \ldots \text{ since } \overline{D}_{\dot\beta}\Phi = 0 \tag{14.228}$$

$$= 2i\overline{\xi}^{\dot\beta}\partial_{\beta\dot\beta}\psi^\beta \tag{14.229}$$

yet again. Clearly these methods are rather clean and powerful. Now it is time to rediscover the invariant Lagrangian of Wess and Zumino. We need the idea of superspace integration [11]. We will start in a one-dimensional case, where

$$f(x,\theta) = f_0(x) + f_1(x)\theta \tag{14.230}$$

and demand linearity, so that

$$\int d\theta\, f = f_0\int d\theta + f_1\int d\theta\,\theta$$
$$= I_0 f_0 + I_1 f_1 \tag{14.231}$$

where I_0 and I_1 are (universal) constants. We then demand translational invariance, so that

$$\begin{aligned}\int d(\theta+\varepsilon)f(x,\theta+\varepsilon) &= \int d\theta\{(f_0+f_1\varepsilon)+f_1\theta\} \\ &= I_0(f_0+f_1\varepsilon)+I_1 f_1 \qquad (14.232)\end{aligned}$$

and have $I_0 \equiv 0$. This means that

$$\int d\theta(f_0+f_1\theta) = f_1 I_1 \qquad (14.233)$$

$$\text{and} \quad \int d\theta \sim \partial/\partial\theta. \qquad (14.234)$$

The integral I_1 is conventionally taken as unity, so that

$$\int d\theta = 0$$

$$\int d\theta\,\theta = 1 \qquad (14.235)$$

$$\text{and} \quad \int d\theta\, f = f_1 \qquad (14.236)$$

$$= \frac{\partial f}{\partial\theta}. \qquad (14.237)$$

Also

$$\begin{aligned}\int d\theta\,\frac{\partial f}{\partial\theta} &= \int d\theta\, f_1 \\ &= 0 \qquad (14.238)\end{aligned}$$

so integration by parts works while

$$\int d\theta\,\theta f = f_0 \qquad (14.239)$$

$$= f(\theta=0) \qquad (14.240)$$

$$\text{so that} \quad \delta(\theta) \equiv \theta. \qquad (14.241)$$

Notice that $\delta(\theta)\delta(\theta) = 0$.

We define

$$\int d^2\theta = \int (-)\varepsilon_{\alpha\beta}d\theta^\alpha d\theta^\beta \qquad (14.242)$$

$$\text{and} \quad \int d^2\overline{\theta} = \int \varepsilon_{\dot{\alpha}\dot{\beta}}d\overline{\theta}^{\dot{\alpha}}d\overline{\theta}^{\dot{\beta}} \qquad (14.243)$$

$$\text{with} \quad \int d^4\theta = \int d^2\theta d^2\overline{\theta}. \qquad (14.244)$$

Now

$$\int d^2\theta\ f(\theta) = \int (-)\varepsilon_{\alpha\beta} d\theta^\alpha d\theta^\beta \{ f_{(0)} - 2\theta^\gamma f_{(1)\gamma} - \theta^2 F_{(2)} \} \tag{14.245}$$

$$= 0 + 0 + f_{(2)}\varepsilon_{\alpha\beta} \int d\theta^\alpha d\theta^\beta \{ -\varepsilon_{\rho\lambda}\theta^\rho\theta^\lambda \} \tag{14.246}$$

$$= (-)4 f_{(2)} \int d\theta^{(1)} d\theta^{(2)} \theta^{(1)} \theta^{(2)} \tag{14.247}$$

$$= 4 f_2. \tag{14.248}$$

But recall

$$\partial^2 f = \partial^\alpha \partial_\alpha f \tag{14.249}$$

$$= 0 + 0 + \varepsilon^{\alpha\beta} \partial_\beta \partial_\alpha \{ (-) f_{(2)} \theta^2 \} \tag{14.250}$$

$$= (-) f_{(2)} \varepsilon^{\alpha\beta} \partial_\beta \{ 2\theta_\alpha \} \tag{14.251}$$

$$= (-)2 f_{(2)} \varepsilon^{\alpha\beta} \varepsilon_{\gamma\alpha} \partial_\beta \theta^\gamma \tag{14.252}$$

$$= 4 f_2 \tag{14.253}$$

also. Hence

$$\int d^2\theta f = \partial^2 f. \tag{14.254}$$

Invariants and Lagrangians

Consider the invariant

$$I = \int d^4x\, d^2\theta\, d^2\overline{\theta}\ \mathcal{L}(\Phi, D_\alpha \Phi, \cdots), \tag{14.255}$$

which is index saturated in the usual way. We avoid Taylor expansion (although it is sometimes easier) by noticing that

$$\left.\begin{aligned} \partial_\alpha &= D_\alpha \\ \overline{\partial}_{\dot\alpha} &= \overline{D}_{\dot\alpha} \end{aligned}\right\} \text{inside} \int d^4x\,, \text{by throwing divergences away.} \tag{14.256}$$

Then

$$I = \int d^4x\ D^2 \overline{D}^2 \mathcal{L} \Big|_0 \tag{14.257}$$

and we evaluate this integral by use of our identities and component definitions. For example, with chiral Φ again,

$$I = \int d^4x\, d^4\theta\, \overline{\Phi}\, \Phi \tag{14.258}$$

$$= \int d^4x\, D^2\, \overline{D}^2\, \overline{\Phi}\, \Phi\Big|_0 \tag{14.259}$$

$$= (-) \int d^4x\, \overline{D}^2 D^2\, \overline{\Phi}\, \Phi\Big|_0 \qquad (\textstyle\int \partial^2 \to 0) \tag{14.260}$$

$$= (-) \int d^4x\, \overline{D}^2\, \overline{\Phi}\, D^2\, \Phi\Big|_0 \qquad (\text{since } D^2\, \overline{\Phi} = 0) \tag{14.261}$$

$$=(-)\int d^4x\, \{\overline{\Phi}\, \overline{D}^2 D^2\, \Phi + 2(\overline{D}_{\dot{\alpha}}\overline{\Phi})\overline{D}^{\dot{\alpha}}\, D^2\, \Phi + (\overline{D}^2\, \overline{\Phi})\, D^2\, \Phi\}\Big|_0 \tag{14.262}$$

$$=(-)\int d^4x\{(-)16\overline{\Phi}\,\Box\Phi + 2(\overline{D}_{\dot{\alpha}}\overline{\Phi})\, 4i\varepsilon^{\dot{\alpha}\dot{\beta}}\partial_{\gamma\dot{\beta}} D^{\gamma}\, \Phi + (\overline{D}^2\overline{\Phi})(D^2\Phi)\}\Big|_0 \tag{14.263}$$

$$=(-)\int d^4x\, \{(-)16\overline{A}\Box A + 8i2\overline{\psi}_{\dot{\alpha}}\varepsilon^{\dot{\alpha}\dot{\beta}}\partial_{\gamma\dot{\beta}}(-)2\psi^{\gamma} + 16\overline{F}\, F\} \tag{14.264}$$

since $\overline{\Phi} = \overline{A} - 2\overline{\psi}_{\dot{\alpha}}\overline{\theta}^{\dot{\alpha}} - \overline{F}\,\overline{\theta}^2 \Rightarrow \overline{D}_{\dot{\alpha}}\overline{\Phi} = 2\overline{\psi}_{\dot{\alpha}}$

$$= (-)16 \int d^4x\{(-)\overline{A}\Box A + 2i\overline{\psi}_{\dot{\alpha}}\partial^{\beta\dot{\alpha}}\psi_{\beta} + \overline{F}\, F\}. \tag{14.265}$$

This is (−32) times what is usually called the *Wess and Zumino model Lagrangian*. Often it is put into the Wess–Zumino language by writing

$$\left.\begin{array}{ccc} A & \to & A + iB \\ F & \to & F - iG \end{array}\right\} \text{and} \quad \psi \leftrightarrow \begin{pmatrix} \psi_{\alpha} \\ \overline{\psi}^{\dot{\alpha}} \end{pmatrix} \tag{14.266}$$

so that

$$\overline{\psi}\, \partial\!\!\!/ \psi = \overline{\psi} \begin{pmatrix} 0 & \sigma^{\mu} \\ \overline{\sigma}^{\mu} & 0 \end{pmatrix} \psi \tag{14.267}$$

$$= 2\overline{\psi}_{\dot{\beta}}\partial^{\lambda\dot{\beta}}\psi_{\lambda} \tag{14.268}$$

we get $$I = (-32) \int d^4x \left\{(-)\frac{1}{2}[(\partial_{\mu}A)^2 + (\partial_{\mu}B)^2] + i\overline{\psi}\, \partial\!\!\!/ \psi + F^2 + G^2\right\} \tag{14.269}$$

= (−32). (Usual form of Wess–Zumino model Lagrangian.)

Notice that when we write invariant Lagrangians like this in supersymmetry, they are really only invariant up to a total derivative which is then thrown away under an integral. This is exactly the way the D-term transformed in the full scalar case, and the F-term transformed in the chiral case.

At any rate we can see that Equation (14.265) contains kinetic-like terms, and we take

$$L_0 = \left(-\frac{1}{32}\right) \int d^4x\, d^4\theta\, (\overline{\Phi}\, \Phi) \tag{14.270}$$

as the first part of our general chiral scalar Lagrangian. We seek mass terms, and interaction terms. Consider

$$I = \int d^4x\, d^4\theta\; P(\Phi) \tag{14.271}$$

where P is polynomial and therefore chiral if Φ is chiral. This is invariant. Indeed $\int d^4x \int d^4\theta\; P = \int d^4x\, d^2\theta\, \overline{D}^2 P \equiv 0$, or more generally, $\int d^4x\, d^4\theta\; \mathcal{Z} = \int d^4x\, d^2\theta (\overline{D}^2 \mathcal{Z})$ where $\mathcal{Z}$ is general but $(\overline{D}^2 \mathcal{Z})$ is chiral because $\overline{D}^3 \equiv 0$. Then

$$I = \int d^4x\; D^2 P\Big|_0 \tag{14.272}$$

$$= \int d^4x\; D^\alpha D_\alpha P\Big|_0 \tag{14.273}$$

$$= \int d^4x\; D^\alpha P' D_\alpha \Phi \tag{14.274}$$

$$= \int d^4x [P''(D_\alpha \Phi)^2 + P' D^2 \Phi]\Big|_0 \tag{14.275}$$

$$= 4 \int d^4x\, [P''(A)\psi^2 + P'(A) F]. \tag{14.276}$$

Now we can see what to do. We write

$$L_m = \left(-\frac{m}{16}\right) \int d^4x\, \left\{ \int d^2\theta\; \Phi^2 + d^2\overline{\theta}\; \overline{\Phi}^2 \right\} \tag{14.277}$$

$$= \left(-\frac{m}{2}\right) \int d^4x\, \{\psi^2 + AF + \overline{\psi}^2 + \overline{A}\overline{F}\} \tag{14.278}$$

then with

$$\left.\begin{array}{ccc} A & \to & A + iB \\ F & \to & F - iG \end{array}\right\}, \psi \to \begin{pmatrix} \psi_\alpha \\ \overline{\psi}^{\dot{\alpha}} \end{pmatrix} \tag{14.279}$$

as before,

$$L_m = (-m) \int d^4x\, \left\{ \frac{1}{2} \overline{\psi}\psi + AF + BG \right\} \tag{14.280}$$

and we have our mass terms.

Finally we need interaction terms (not too high in powers of scalars, or we face nonrenormalizability), and we try

$$L_{int} = \left(-\frac{g}{24}\right)\int d^4x\,\{D^2\Phi^3 + \overline{D}^2\overline{\Phi}^3\} \tag{14.281}$$

$$= \left(-\frac{g}{24}\right)\int d^4x\,4\{6A\psi^2 + 3A^2F + 6\overline{A}\,\overline{\psi}^2 + 3\overline{A}^2\overline{F}\} \tag{14.282}$$

$$= (-g)\int d^4x\,\{\overline{\psi}(A + B\gamma_5)\psi + (A^2 - B^2)F + 2ABG\} \tag{14.283}$$

when we make the, by now familiar, substitutions.

Notes

1.

$$\frac{1}{4}\int d^4x\,d^2\theta = \frac{1}{4}\int d^4x\,D^2 \tag{14.284}$$

picks out the F-component of a chiral integrand.

2.

$$D^2\overline{D}^2\Phi = D^\alpha D_\alpha \overline{D}^2\Phi \tag{14.285}$$

$$= D^\alpha\{\overline{D}^2 D_\alpha - 4i\partial_{\alpha\dot{\beta}}\overline{D}^{\dot{\beta}}\}\Phi \tag{14.286}$$

$$= D^\alpha\overline{D}^2 D_\alpha\Phi + \left\{0 \text{ under } \int d^4x\right\} \tag{14.287}$$

$$\text{so}\quad \frac{1}{16}\int d^4x\,d^4\theta = \frac{1}{16}\int d^4x\,d^2\overline{D}^2 \tag{14.288}$$

$$= \frac{1}{16}\int d^4x\,D^\alpha\overline{D}^2 D_\alpha \tag{14.289}$$

picks out the D-component of an integrand. This latter form is particularly heavily used, as it gives a convenient order of terms when dealing with gauge invariance.

3. $\mathcal{Z}$ general $\Rightarrow$ $\overline{D}^2\mathcal{Z}$ is chiral.
4. A product of two chiral fields is again chiral, by the Leibnitz rule. To find the components

$$A_3 = \mathcal{A}_1 \cdot \mathcal{A}_2|_0 \tag{14.290}$$

$$= A_1 A_2 \tag{14.291}$$

then the real form is

$$A_3 = A_1 A_2 - B_1 B_2 \tag{14.292}$$

$$B_3 = A_1 B_2 - B_1 A_2 \tag{14.293}$$

$$\psi_{(3)\alpha} = (-)\,\frac{1}{2} D_\alpha \mathcal{A}_1 \cdot \mathcal{A}_2\Big|_0 \tag{14.294}$$

$$= (-)\,\frac{1}{2}(D_\alpha \mathcal{A}_1)\mathcal{A}_2 + \mathcal{A}_1\left(-\frac{1}{2}D_\alpha \mathcal{A}_2\right)\Big|_0 \tag{14.295}$$

$$= \psi_{(1)\alpha} A_2 + A_1 \psi_{(2)\alpha} \tag{14.296}$$

$$\psi_{(3)} = (A_1 - \gamma_5 B_1)\psi_{(2)} + (A_2 - \gamma_5 B_2)\psi_{(1)}. \tag{14.297}$$

Superpotential

There may, of course, be several chiral scalars ψ_i and we can now write the general Lagrangian for chiral superfields. There is a kinetic term obtained simply by summing over the fields

$$L_0 = \sum_i \left(-\frac{1}{32}\right) \int d^4x\, d^4\theta\, (\overline{\Phi}_i \Phi_i). \tag{14.298}$$

Then there are mass terms, and interaction terms, which are generally grouped together into what is called the *superpotential* $W(\Phi)$.

In our language we have the extra part of the Lagrangian (with m_{ij} and g_{ijk} real and symmetric) as

$$\left(-\frac{1}{8}\right) \int d^4x\, d^2\theta\; W(\Phi) + h.c. \tag{14.299}$$

$$\text{where}\quad W(\Phi) = \frac{1}{2} m_{ij} \Phi_i \Phi_j + \frac{1}{3} g_{ijk} \Phi_i \Phi_j \Phi_k. \tag{14.300}$$

Writing ϕ where we previously wrote A, to standardize the notation, the whole thing is described by

$$\mathcal{L} = \partial_\mu \phi_i^\dagger \partial^\mu \phi_i + 2i\overline{\psi}_i \overline{\sigma}^\mu \partial_\mu \psi_i + F_i^\dagger F_i + m_{ij}\phi_i F_j$$
$$- \frac{1}{2} m_{ij} \psi_i \psi_j + g_{ijk}\phi_i \phi_j F_k - g_{ijk} \psi_i \psi_j \phi_k + h.c.. \tag{14.301}$$

The Euler–Lagrange field equations from this Lagrangian are

$$\partial_\mu \partial^\mu \phi_i = m_{ij} F_j^\dagger + 2g_{ijk}\phi_j^\dagger F_k^\dagger \tag{14.302}$$
$$i\overline{\sigma}^\mu \partial_\mu \psi_i = -m_{ij}\overline{\psi}_j - 2g_{ijk}\overline{\psi}_j \phi_k^\dagger \tag{14.303}$$
$$F_i^\dagger = -m_{ij}\phi_j - g_{ijk}\phi_j\phi_k \tag{14.304}$$
$$= -\frac{\partial W(\phi)}{\partial \phi_i} \tag{14.305}$$

where $W(\phi)$ is the superpotential with $\Phi_i \rightarrow \phi_i$. The last equation of motion, Equation (14.306), confirms that the F_i are auxiliary fields—no derivatives appear—and they may be algebraically removed. Then

$$\mathcal{L} = \partial_\mu \phi_i^\dagger \partial^\mu \phi_i + i\overline{\psi}_i \overline{\sigma}^\mu \partial_\mu \psi_i - F_i^\dagger F_i - \left(\frac{1}{2}m_{ij}\psi_i\psi_j + g_{ijk}\psi_i\psi_j\phi_k + h.c.\right) \tag{14.306}$$
$$= \partial_\mu \phi_i^\dagger \partial^\mu \phi_i + i\overline{\psi}_i \overline{\sigma}^\mu \partial_\mu \psi_i - |m_{ij}\phi_j + g_{ijk}\phi_j\phi_k|^2 - \left(\frac{1}{2}m_{ij}\psi_i\psi_j + G_{ijk}\psi_i\psi_j\phi_k + h.c.\right). \tag{14.307}$$

Notice that the tree level effective potential is

$$\mathcal{V} = F_i^\dagger F_i \tag{14.308}$$
$$= |F_i|^2 \tag{14.309}$$

which is always ≥ 0. Absolute minima of $\mathcal{V}$ are where $F_i \equiv 0$. Switching back to the real form (with Majorana fermions) this Lagrangian reads

$$\mathcal{L} = \partial_\mu \phi_i^\dagger \partial^\mu \phi_i^\dagger - \frac{i}{2}\overline{\Psi}_i \not\partial \Psi_i - \frac{1}{2}m_{ij}\overline{\Psi}_i \Psi_j - \frac{g_{ijk}}{2}[\overline{\Psi}_i\Psi_j - \overline{\Psi}_i\gamma_5\Psi_j]\phi_k - \frac{g_{ijk}}{2}[\overline{\Psi}_i\Psi_j + \overline{\Psi}_i\gamma_5\Psi_j]\phi_k^\dagger \tag{14.310}$$

which most phenomenologists prefer, since they are happier with Dirac γ-matrices and not with dotted and undotted Weyl spinors. The Feynman rules are normally given in this form.

Despite many opportunities supersymmetry ignored them all, and more or less was dragged protesting into our four-dimensional world in 1974 by Wess and Zumino [1, 2, 9] who were referring to a paper by Ramond [3] aiming to introduce particles with half-integer spin into a string theory and also referring to a paper by Neveu and Schwarz [4] suggesting the addition of d fermionic field doublets. At any rate Wess and Zumino constructed several supersymmetric models. The simplest involved a single Majorana (self-charge-conjugate Dirac) field (ψ), a pair of real scalar and pseudoscalar boson fields (A and B), and a pair of real scalar and pseudoscalar bosonic auxiliary fields (F and G). This was invariant under the infinitesimal transformations (the notation has changed over subsequent years).

$$\delta A = \overline{\alpha}\psi \tag{14.311}$$
$$\delta B = (-)i\overline{\alpha}\gamma_5\psi, \tag{14.312}$$

which are connected to

$$\delta\psi = \partial_\mu(A + i\gamma_5 B)\gamma^\mu\alpha + (F - i\gamma_5 G)\alpha, \tag{14.313}$$
$$\delta F = \overline{\alpha}\gamma^\mu\partial_\mu\psi, \tag{14.314}$$
$$\delta G = (-)i\overline{\alpha}\gamma_5\gamma^\mu\partial_\mu\psi \tag{14.315}$$

where α is an arbitrary constant infinitesimal Majorama fermion c-number parameter. The most general real Lorentz-invariant, parity-conserving, renormalizable Lagrangian density built from these objects is

$$\mathcal{L} = (-)\frac{1}{2}\partial_\mu\mathbf{A}\partial^\mu\mathbf{A} - \frac{1}{2}\partial_\mu\mathbf{B}\partial^\mu\mathbf{B} - \frac{1}{2}\overline{\psi}\gamma^\mu\partial_\mu\psi + \frac{1}{2}(\mathbf{F}^2 + \mathbf{G}^2) + m\left(\mathbf{FA} + \mathbf{GB} - \frac{1}{2}\overline{\psi}\psi\right) + g\left[\mathbf{F}(\mathbf{A}^2 + \mathbf{B}^2) + 2\mathbf{GAB} - \overline{\psi}(\mathbf{A} + i\gamma_5\mathbf{B})\psi\right]. \tag{14.316}$$

Since the auxiliary fields **F** and **G** enter quadratically, we can derive an equivalent Lagrangian by setting them equal to the values given by the field equations to see that

$$\mathbf{F} = (-)m\mathbf{A} - g(\mathbf{A}^2 + \mathbf{B}^2) \tag{14.317}$$
$$\mathbf{G} = (-)m\mathbf{B} - 2g\mathbf{AB}. \tag{14.318}$$

The Lagrangian density then becomes

$$\mathcal{L} = (-)\frac{1}{2}\partial_\mu A\partial^\mu A - \frac{1}{2}\partial_\mu B\partial^\mu B - \frac{1}{2}\overline{\psi}\gamma^\mu\partial_\mu\psi + \frac{1}{2}(F^2 + G^2) + m\left[FA + GB - \frac{1}{2}\overline{\psi}\psi\right] + g\left[F(A^2 + B^2) + 2GAB - \overline{\psi}(A + i\gamma_5 B)\psi\right]. \tag{14.319}$$

Since the auxiliary fields F and G enter quadratically, we can derive an equivalent Lagrangian by setting them equal to the values given by the field equations

$$F = -mA - g(A^2 + B^2), \tag{14.320}$$
$$G = -mB - 2gAB. \tag{14.321}$$

The Lagrangian density then becomes

$$\mathcal{L} = \left[-\frac{1}{2}\partial_\mu A\partial^\mu A - \frac{1}{2}\partial_\mu B\partial^\mu B - \frac{1}{2}\overline{\psi}\gamma^\mu\partial_\mu\psi - \frac{1}{2}m^2(A^2 + B^2) - \frac{1}{2}m\overline{\psi}\psi - gmA(A^2 + B^2) - \frac{1}{2}g^2(A^2 + B^2)^2 - g\overline{\psi}(A + i\gamma_5 B)\psi\right]. \tag{14.322}$$

This Lagrangian density exhibits relations not only between scalar and fermion masses, but also between Yukawa interactions and scalar self-couplings, which are characteristic of supersymmetric theories.

Unknown to Wess and Zumino, at the time of their first papers on supersymmetry in four space–time dimensions, this symmetry had already appeared in a pair of papers published in the Soviet Union. In 1971 Gol'fand and Likhtman [5] had extended the algebra of the Poincaré group to a superalgebra and used the requirement of invariance under this superalgebra to construct supersymmetric field theories in four space–time dimensions. Independently, Volkov and Akulov [6] in 1973 discovered what we would call spontaneously broken supersymmetry, but they used their formalism to identify the Goldstone fermion associated with supersymmetry breaking with a neutrino.

Coleman and Mandula [7] proved a celebrated theorem to the effect that the only Lie algebra (as opposed to superalgebra) of symmetry generators consists of the generators P_μ and $J_{\mu\nu}$ of translations and homogeneous Lorentz transformations, together with possible internal symmetry generators, which commute with P_μ and $J_{\mu\nu}$ and act on physical states by multiplying them with spin-independent, momentum-independent Hermitian matrices. There are a number of extra conditions that are not needed for our purposes, a generalization to include the remaining Lie algebra of the conformal group by Haag, Lopuszanski, and Sohnius [8]. This publication effectively establishes the result using the elementary particle states, both massive and massless, known at that time.

At this state of our understanding we do not need to know how to construct all massive and massless representations of supersymmetry. It is worth knowing, however, that there is only one kind of massless supermultiplet in theories with simple supersymmetry. There are no massless particles that are not accompanied by a superpartner or ones that have more than one superpartner. How can the known quarks, leptons, and gauge bosons fit into this picture? We assume that the supersymmetry generator commutes with the generators of the $SU(3) \times SU(2) \times U(1)$ gauge group. The quarks and leptons cannot be in the same multiplets as the gauge bosons. In the limit of high energy where $SU(2) \times U(1)$ symmetry breaking can be neglected, the massless quarks and leptons of each color and flavor are in supermultiplets with pairs of massless squarks and sleptons of zero helicity and the same color and flavor, while the massless gauge bosons are accompanied by massless gaugines of helicity $\pm\frac{1}{2}$ comprising an adjoint representation of $SU(3) \times SU(2) \times U(1)$.

Because gravity exists we know that there must exist a massless particle of helicity ± 2, the *graviton*. There are no conserved quantities to which a soft massless particle with $|\lambda| > 2$ could couple. We conclude that the graviton must be in a supermultiplet with a massless particle of helicity $\pm\frac{3}{2}$. This is a gravitino, coupled to the supersymmetry generators themselves. The field theory of this multiplet is known as supergravity.

References

1. J. Wess and B. Zumino, *Nucl. Phys.* B70 (1974): 39.
2. J. Wess and B. Zumino, *Phys. Lett.* 49B (1974): 52.
3. P. Raymond, *Phys. Rev.* D3 (1971): 2415.
4. A. Neveu and J.H. Schwarz, *Nucl. Phys.* B31 (1971): 86; *Phys. Rev.* D4 (1971): 1109.
5. Yu.A. Gol'fand and E.P. Likhtman, *JETP Lett.* 13 (1971): 323.
6. D.V. Volkov and V.P. Akulov, *Phys. Lett.* 13 (1971): 323.
7. S. Coleman and J. Mandula, *Phys. Rev.* 159 (1967): 1251.
8. R. Haag, J.T. Lopuszanski, and M. Sohnius, *Nucl. Phys.* B88 (1975): 257.
9. J. Wess and B. Zumino, *Nucl. Phys.* B78 (1974): 1.
10. C. Callan, S. Coleman, J. Wess, and B. Zumino, *Phys. Rev.* 177 (1969): 2247.
11. F.A. Berezin, *The Method of Second Quantization*. Academic Press, New York, 1966.

Problems

14.1 Confirm contact with previous notation by retrieving $x_i' = x_i + \varepsilon_{ijk}\theta_j x_k$ for [3] rotations.

14.2 Repeat Problem 14.1 but for $\theta_\alpha' = \theta_\alpha - \frac{i}{2}(\theta_i \sigma_i)_\alpha^\beta \theta_\beta$ lowering the spinor index.

14.3 Check the reality of δx^μ.

14.4 Retrieve: $L_i \Rightarrow \varepsilon_{ijk} x_j(-i\partial_k) + \frac{1}{2}(\sigma_i)_\beta^{\ \alpha}\theta^\beta \partial_\alpha - \frac{1}{2}(\overline{\sigma}_i)^{\dot{\alpha}}_{\ \dot{\beta}}\overline{\theta}^{\dot{\beta}}\overline{\partial}_{\dot{\alpha}}$ and check directly that $[(-)i\vec{\theta} \cdot \vec{L}]\theta_\alpha = \frac{i}{2}(\sigma_i)_\alpha^\beta \theta_i \theta_\beta$ as it should be to induce transformations of fields as imposed. Note that this exercise is supposed to draw attention to $\theta_i\ V^s\ \theta_\alpha$, and to remind us that in our notation $C_{\alpha\beta} = \varepsilon_{\alpha\beta}$, but $(C^{-1})^{\alpha\beta} = (-)\varepsilon^{\alpha\beta}$, so that $(\sigma_i)_\alpha^{\ \beta} = (-)C_{\alpha\alpha'}(\sigma_i)_{\beta'}^{\ \alpha'}(C^{-1})^{\beta'\beta}$].

14.5 Check Q_α and $\overline{Q}_{\dot{\alpha}}$ are related by conjugation. (Watch $[\sigma_{\alpha\dot{\beta}}]^* = \sigma_{\beta\dot{\alpha}}$, but $\overline{\partial}_\mu = (-)\partial_\mu \Rightarrow \overline{(\partial_{\alpha\dot{\beta}})} = (-)\partial_{\beta\dot{\alpha}}$.)

14.6 Check $\{Q_\alpha, \overline{Q}_{\dot{\beta}}\} = 2\sigma_{\alpha\dot{\beta}}P_\mu$ directly. (Warning: Watch out for Wess and Zumino here.)

14.7 Check the statement on subgroup action.

14.8 Work through our Euclidean example. Confirm that rotations are unchanged, but that for translations $\vec{x} \to \vec{x} - \vec{a}$ under right action compared to $\vec{x} \to \vec{x} + \vec{a}$ under left action.

14.9 Check $\{D_\alpha, Q_\beta\} = 0$ directly.

14.10 (Trivial) Show D_α and $\overline{D}_{\dot{\alpha}}$ give Leibnitz rules, then work out $D_\alpha \overline{D}_{\dot{\beta}} \theta_\gamma \overline{\theta}_{\dot{\delta}}$ a few different ways for practice.

14.11 Work out the $\overline{\theta}$ terms and check consistency.

14.12 Check

$$
\begin{aligned}
D^2\Phi|_0 &= D^\alpha D_\alpha \Phi\big|_0 \\
&= D^\alpha D_\alpha(-F\theta^2)\big|_0 \\
&= D^\alpha(-2F\theta_\alpha)\big|_0 \\
&= 2D_\alpha(F\theta^\alpha)\big|_0 \\
&= 4F.
\end{aligned}
$$

Notice that, if we go on, we get

$$
\begin{aligned}
\overline{D}_{\dot\alpha} D_\alpha \Phi\big|_0 &= \{D_\alpha, \overline{D}_{\dot\alpha}\}\Phi| \qquad (\text{since } \overline{D}_{\dot\alpha}\Phi \equiv 0) \\
&= (-)2i\partial_{\alpha\dot\alpha}\Phi|_0 \\
&= -2i\partial_{\alpha\dot\alpha} A.
\end{aligned}
$$

and the differential constraints are now automatic.

14.13 Check the result of Problem 14.12 explicitly using the previous (explicit) form of Φ.

14.14 Check through the other pieces—obviously $\overline{D}^3 = 0 = D^3$, so this stops—in agreement with our other expanded form.

14.15 For example (work through)

$$
\begin{aligned}
\int d^2\overline{\theta}\,\overline{f} &= 4\overline{f}_{(2)} \\
&= \overline{\partial}^2 f.
\end{aligned}
$$

14.16 Work out the things that follow by this method

$$
\begin{aligned}
F_3 &= F_1A_2 + A_1F_2 - \psi_{(1)\alpha}\psi^\alpha_{(2)} + \psi^\alpha_{(1)}\psi_{(2)\alpha} \\
F_3 &= F_1A_2 + F_2A_1 + G_1B_2 + B_1G_2 + \overline{\psi}_{(1)}\psi_{(2)} \\
G_3 &= G_1A_2 + G_2A_1 - F_1B_2 - F_2B_1 - \overline{\psi}_{(1)}\gamma_5\psi_{(2)}.
\end{aligned}
$$

14.17 Notice that the Coleman–Mandula theorem deals only with transformations that take bosons into bosons and fermions into fermions and are therefore generated by operators that satisfy commutation relations rather than anticommutation relations. This raises the question of whether a relativistic theory can have symmetries taking fermions and bosons into each other and therefore satisfy anticommutation relations rather than commutation relations. Show that supersymmetry is the only possible solution to this situation.

14.18 Show that the most general symmetry algebra allowed under the assumptions of the Coleman–Mandula theorem in the case where all particles are massless consists of internal symmetry generators plus either the Poincaré algebra or the conformal algebra.

14.19 Calculate the change in the Wess and Zumino Lagrangian density under the space–time supersymmetry transformation.

14.20 Find a set of 2×2 matrices that form a graded Lie algebra containing fermionic as well as bosonic generators.

14.21 In 2 space and 1 time dimensions you can take the generators of the Lorentz group as $A_1 = (-)iJ_{10}$, $A_2 = (-)iJ_{20}$, and $A_3 = J_{12}$.
The commutation relations of the Poincaré algebra are $[A_i, A_j] \equiv i\sum_k \varepsilon_{ijk} A_k$ so the representations of the homogeneous Lorentz group in $2+1$ space–time dimensions are labelled with a single positive integer or half-integer A. Following the approach of Haag, Lopuszanski, and Sohnius derive the most general symmetry that can be sustained.

Index

N

O

P

Printed in the United States
by Baker & Taylor Publisher Services